Backyard Poultry Raising

Other books by John F. Adams

AN ESSAY ON BREWING, VINTAGE AND DISTILLATION,
TOGETHER WITH SELECTED REMEDIES FOR HANGOVER MELANCHOLIA

BEEKEEPING: THE GENTLE CRAFT

TWO PLUS TWO EQUALS MINUS SEVEN

Backyard Poultry Raising

*The Chicken Growing, Egg Laying,
Feather Plucking, Incubating, Caponizing,
Finger Licking Handbook*

by JOHN F. ADAMS
with illustrations by the author

DOUBLEDAY & COMPANY, INC., GARDEN CITY, NEW YORK, 1977

Library of Congress Cataloging in Publication Data

Adams, John Festus, 1930–
Backyard poultry raising.

Includes index.
1. Poultry. I. Title.
SF487.A33   636.5
ISBN: 0-385-11509-1
Library of Congress Catalog Card Number 76–23746

Copyright © 1977 by John F. Adams
All Rights Reserved
Printed in the United States of America

FOR PEARL U. ADAMS, my mother, from whom I learned everything of practical worth I know about chickens.

CONTENTS

THE BIG, FAT BUTTER-AND-EGG MAN	1
THE WHOLESOME CHICKEN	9
THE EGG AND HOW TO GET IT	33
THE CHICKEN AT LARGE	49
OTHER POULTRY, EXCLUDING RABBITS	63
CHICKENS CANTANKEROUS AND DROOPY	81
HOUSING THE HUMBLE CHICKEN	91
THE FEASIBLE CHICKEN	105
WHEN A SPARROW FALLS	111
GLOSSARY	125
INDEX	129

List of Illustrations

COMMERCIALLY MADE BROODER	13
HOMEMADE BROODER	14
HOMEMADE FEEDER FOR BABY CHICKS	17
TYPICAL WATERER	18
SIMPLE PLYWOOD SHELTER	19
WOODEN FEEDER	25
TWO-COMPARTMENT HOPPER	38
OVERHEAD FEEDER	40
SIMPLE A-FRAME COOP	56
OPEN SHELTER	93
MY CHICKEN HOUSE	94
DROPPINGS PIT	96
HINGED-DOOR NEST	98
SIMPLE NEST	99
FREESTANDING FLY-UP FEEDER	100
SUN PORCH FOR CHICKENS	101

The Big, Fat Butter-and-Egg Man

The second dullest man I have ever known was an army corporal who spent his career in charge of the garbage pit at an army post in Virginia. The *dullest* man I have ever met was a professor of poultry science at a state college in the East. There's something about taking chickens too seriously that robs the mind of an essential particle, some mystical juice or ichor, until the entire personality atrophies. Actually taking any pastime or preoccupation too seriously tends to induce similar effects. A biographical profile I read some years ago chronicled the life and opinions of a cheese merchant. At one point he was quoted as saying that cheese must be the most interesting thing in the world; everywhere he went he found people talking of nothing else. Substitute the Ancient Mariner for the cheese merchant, and internal parasites of the chickens for cheese, and you have a pretty good image of the poultry professor I used to know.

Take chickens, then, but don't take them too seriously. Keeping a clear-eyed perspective on why you want to get into chickens will help preserve that vision. The reasons any particular person keeps, or wants to keep, chickens will be found to include one or more of these three options: for eggs, for meat, or for the hell of it.

Someone who wants to keep chickens for their eggs is probably motivated by two interests—quality and economy. Eggs were among the first commodities to suffer the quality erosion associated with mass production and mass distribution of foods. An egg provides the perfect container to conceal shoddy merchandise. Even hot dogs and sausages are grudgingly being constrained by law to confess on their labels that they consist primarily of hog maws and pig snouts, lips, and ears, rather than the devious caveat "specialty meat" they were permitted only a few years ago. But the ways of determining what's inside an egg are limited. Candling—visually inspecting it against a strong light—will show if it is fertile or contains a blood clot, which is harmless. (Some say fertile eggs are more wholesome than infertile eggs.) Candling can also show if it has more than one yolk, or is spoiled. But the flavor of the egg was concocted and determined in the dark insides of the hen and is chiefly influenced by what was put into the hen by way of food and by the physical circumstances of her life. These elements, plus the egg's relative degree of freshness, are secrets only he who eats the egg knows for sure. No "truth in labeling law" can force the revelation of a clue to these wholesome secrets before the shell of the egg is cracked. And the word "fresh" can be more devious, if less sinister, than the designation "specialty meats."

The "fresh" egg of commerce is, practically speaking, simply an unrotten egg. Eggs, even without any kind of refrigeration short of protecting them from heat, will keep for weeks. They must; the

The Big, Fat Butter-and-Egg Man

larger nesting birds, including the chicken, may take at least a couple of weeks to accumulate enough eggs to be worth their while to incubate. An exception which perhaps proves the rule is an ostrichlike bird of the Argentine Pampas. These birds, the greater rhea and the Darwin's rhea, are in their reproductive habits "polygamous-polyandrous," meaning they practice reversed sexual roles with multiple mates. The weather in the region of their occurrence is so hot over the nesting season that an egg would spoil in a day or so if it was not incubated and "chilled" down to body temperature. Consequently, coaxed by the male, a flock of hens will each lay *one* egg in a nest he has prepared. *He*, then, incubates the clutch of eggs provided for him in a single day, from several to nearly twenty eggs, and the hens go on to lay in the nests of some dozen or twenty more husbands on successive days. These birds fascinated Charles Darwin and, reputedly, he perceived the distinction between the two species while eating a drumstick of the slightly smaller variety, which has ever since carried the name of the great naturalist.

But generally speaking, eggs keep a long time. Not always well, but they still keep, for marketing purposes, *fresh*. Probably few people ever confront a really *fresh* egg. Those who have would probably be unanimous in agreeing that fresh eggs are, to the eggs of commerce, as pig lips are to pork chops.

The expression "farm fresh" conjures upon the palate other intimations of quality and flavor, taste images which are reinforced by visual images. Farm fresh suggests the picture of chickens at liberty chasing bugs and relaxing in dust baths, and the aproned farm wife collecting the eggs in a blue enamel kettle from nests lined with sweet-smelling golden straw. Eggs advertised as "farm fresh" will actually deliver the taste promised by that imagery only if you know the seller and the pedigree of his eggs. Otherwise, in commerce the expression or label is about as meaningful as "sweet creamery butter" printed on a carton you pick up in the grocery case next to the oleo. To deserve the label, farm eggs should be not only fresh, but the chicken should have been fed a diet that does not reduce her function to that of a machine processing the least expensive nourishment into the maximum number of inscrutable, unlabelable packages of protein and carbohydrate.

When broken into a pan, a *fresh* egg will reveal a yolk tall and perky as a young breast, almost a perfect hemisphere, surrounded closely and tightly by a white that spreads no farther from the yolk than its own diameter. An *unfresh* egg, often the common fresh egg of commerce, will expose a yolk that immediately flattens away from the hemispherical, around which the white puddles thinly away. The less fresh the egg, the flatter will be the yolk, the easier the yolk will break, and the more watery the white will appear.

The genuine farm-fresh egg will show, when broken, all of the above features of freshness. In addition, the yolk should present not a pale yellow color, but a much more pigmented yellow, sometimes even resembling the color of a hunter's moon just as it crests above the horizon. *Yellow* yellow; saturate orange. The diet of the hen is the principal cause of variance in color. The poultry mix fed commercially will consistently produce the pale yellow yolks. The rougher—I hesitate to use the word because it has come to be suggestive of self-righteous and overgeneralized prejudices—but more *natural* feed of the farm chicken adds color to the eggs. Chickens that have access to greenery will lay eggs with darker colored yolks. The majority of people who try them prefer them hands down to lighter eggs. There is a single exception to the preferability of freshness in an egg. Hard-boiled, a fresh egg tends to resist being peeled, and it is very difficult to remove the shell without also removing about an eighth of the egg.

It is not only the color of the yolk, however, that distinguishes the farm-fresh egg. The whole point of it all is, of course, flavor, and the indications of freshness and the color of the yolk are nothing but promises that eating must judge. I can think of no way to describe the flavor of an egg, so I am helpless in trying to compare it with the flavor of a better egg. Anyone who demands proof of this point, that the farm-fresh egg is not only better, but immensely better, will never trouble himself with raising chickens for their fruit. And barring privileged associations with producers who pro-

The Big, Fat Butter-and-Egg Man

vide you access to farm-fresh eggs—using the expression accurately and not commercially—the only way to have them is to raise them yourself.

Probably most people interested in raising chickens do not need to be convinced of the superiority of the eggs he might produce. Likely, however, he will be equally convinced that growing his own, the eggs will be not only better, but cheaper as well. This may be true, but it depends entirely upon circumstances and situation. If your circumstances require that all feed must be purchased from feed stores, in the long run the eggs will probably not be cheaper. They may be, particularly if you keep a modest-sized flock, but probably over the course of a year the savings will be rather nominal for the number of eggs an average family eats. It should also be remembered that if the chickens are fed exactly the same feed as those grown commercially, the eggs probably will taste about the same as commercially produced eggs. They should be better simply because of their freshness, but ultimately they will probably lack the distinct individuality of farm-fresh eggs. But the amateur's situation might well allow him to provide supplemental feed that costs him little or nothing. Generally speaking, the short cuts which the amateur might employ, taken in the interest of cutting cost, are expedients which also tend to improve the quality of the eggs. The quantity of eggs may be reduced in comparison to the quantity produced by hens fed exclusively on commercially prepared laying mixtures, but they are most likely going to be better eggs. And if a person today goes to the trouble to raise chickens for their eggs, it is likely that he is primarily interested in quality. The extra efforts his methods may put him to will probably be agreeable and part of the general fun of the hobby.

The same principles which relate the diet of the chicken to the flavor of the eggs are equally true if a hobbyist is raising chickens for their meat. Old-timers, and not so old-timers, complain that chicken doesn't taste the way it used to. It doesn't, and the first home-grown chicken you eat will convince you of that, if convincing is necessary. Many people with a health-food orientation are convinced it is the vaccination, medication, and lack of organically grown food in a chicken's diet that are responsible for this degeneration of flavor. Perhaps there is something to these contentions. At any rate, there should be little doubt that diet must be a primary factor in the way the bird finally tastes. Like us, a chicken is what he eats. I believe also that birds raised in the tightly restricted quarters of a "broiler factory" suffer in flavor from lack of exercise and lack of sunshine and air as well as because of their restrictive diet. A chicken that has never eaten a bug cannot be considered a fulfilled chicken.

The age at which the chicken is butchered is also a very significant factor in the way he tastes. A fryer, to be a fryer, must be tender, and of course the younger it is the more tender it will be. Modern commercially raised chickens have been bred into such efficient converters of feed to flesh that the bird may be a two-and-a-half-pound fryer at six weeks. The same bird, if a cockerel, would weigh perhaps ten pounds when it was old enough to crow. When you see a slender little fryer at the butcher's counter, in reality it is hardly more than a baby chick.

My personal preference is to keep fryers until the age of at least ten weeks, at which time it should weigh about four pounds. Also, I always allow a number of birds to mature to serve as roasting chickens. For my taste, the best of all chickens is one that has reached a nearly mature size, six or eight pounds for a hen of a large breed, slightly heavier for a cockerel. Keeping the chickens that long, of course, increases the cost, and after a point the cost of feed begins to gallop away from the retail value of the bird. But again, most people who raise chickens for meat on a small scale are more interested in quality than strict economy. And in the fun of it all.

The ultimate and final reason people raise chickens is simple attachment for birds and animals, but chickens in particular. And chickens are not only plain, they may be fancy as well. For someone who has never gone through the poultry barns of a county fair, not to mention visiting a chicken fanciers' convention, the varieties of exotic chickens are unexpected if not actually unbelievable.

The Big, Fat Butter-and-Egg Man

There are varieties bred for their topknots, for their color, for their feathered legs, for their bare legs, for their size, and even for the color of the eggs they lay. And fancy chickens command fancy prices. They are also often difficult to raise, difficult to breed, lack resistance, and are sometimes contrary and cantankerous—all qualities that endear them to their fanciers. Even the common garden varieties of chicken have a quality of appeal that I respond to for the same reasons I find most birds appealing. I'm touched by the dependency of chickens and by the fact that they are birds and make friends with me. Even their occasionally maddening simpleness is somehow appealing. And despite the passing of time, I have never lost the sense of pleasure of discovery in the collecting of eggs, or of the tactile sensation of holding an egg, particularly a still warm egg, in my hand.

Although a drone to production, the most flashy and eye-catching chicken in the coop is a rooster. Hens find their attentions refreshing but, of course, will lay as many eggs without their attention as they will with their nearly ceaseless romancing. In commerce, the roosters (except breeding stock) never survive pubescence. Unless they prefer fertile eggs, many back-lot chicken ranchers also do not keep roosters (or a rooster) for two perfectly sensible reasons. One, he eats without providing any visible return. Two, he crows early in the morning and, as the mood strikes him, periodically throughout the day as well. I find the sound of a rooster crowing like the chiming of a fine old grandfather clock. For the most part, you don't hear it. When you do notice it, you listen with appreciation. The country would not be the country without the sound of roosters crowing. He also has other qualities to recommend him. In most instances, the rooster is an exceptionally beautiful creature. In addition, he has a self-assurance and confidence which is attractive in the same way these qualities are attractive in wild animals. Even the visible virility of the rooster may be a part of his appeal. With his astonishing promiscuity, it is little wonder that the rooster figures so prominently in mythologies and religions. The great Roman physician Galen observed: *"Triste est omne animal poist coitium, praeter mulierem et gallumque.*" Which is to say, "All animals experience a sadness after coition except the human female and the rooster." Perhaps more gallantly, another Latin authority, probably a medieval monk, makes essentially the same observation, but for the human female substitutes the horse.

The Wholesome Chicken

An old vaudeville gag had it that the chicken must be the world's most useful animal—you can eat it before it's born and after it's dead. For the prospective chicken grower, however, it is almost exclusively the period between usefulnesses which requires his attention. The complete novice might first unravel the conundrum of the even older gag: Which came first, the chicken or the egg?

For the beginner, certainly the chicken comes first. Incubating eggs, other than with special equipment, requires a setting hen—in which case the chicken still comes first. The chicks which he begins with may be only one day old, but the chicken still comes first. The question that has to be met before even considering whether to raise chickens for meat or for eggs, put practically and simply, is, Do you have enough space available to raise chickens at all?

A person actually living inside a town of any size is going to find that he is mostly out of luck regardless of his space. Town and city ordinances established when wayward flocks and peremptory roosters were a real environmental problem took care to prohibit poultry. Not arguing relative merits of domestic animals, it is hard to see why a small number of chickens constitutes more urban nuisance than dogs or cats. Scarcely a residential block is without its nocturnal dog barking away the night in mindless monotony, or a wailing lovelorn cat silhouetting itself on a patio fence. I hasten to add that I like both dogs and cats and have shared space with both species almost without interruption throughout my life. Just making a point: the relative nuisance distinction seems so fine as to be arbitrary or absurd. The laws probably can't be changed, but perhaps conscience can stretch to overlook them on occasion.

While not advocating breaking the law if your particular area flatly prohibits chickens, chickens for broilers or fryers can be raised so unobtrusively that hardly the next-door neighbor need be the wiser. It would be possible, of course, to raise them entirely in a basement, on an inclosed porch, etc., and, short of invasion with a search warrant, no one could be the wiser as long as you keep things tidy. Before they were old enough to become noisy, it would all be over. As a matter of fact, home-sized cage arrangements are available for almost precisely that sort of circumstance. Indeed, raising them in such quarters would actually be almost identical to the way numerous chickens are raised for the commercial market. Except for the rather minor private pleasure of knowing you might be successfully breaking the law, the system would have to be considered self-defeating. You might raise them somewhat cheaper than you could buy them, and eat them fresher, but, in quality, what you produced by that system would be essentially no improvement over the fryers in your supermarket.

If you are limited by an urban environment but choose never-

The Wholesome Chicken

theless to risk confrontation with the local ordinances, by all means test out the immediate neighbors: they are the most likely to find out and the most probable to turn chicken stoolies. Better still, enlist them in the project in a community conspiracy on a cost- and labor-sharing basis. They might even share some space. If conspiracy is often the intangible cement that binds the governers against the governed, it can also work the other way around. Ideally, of course, the beginning chicken rancher, regardless of his prospective scale of operation, should have at least a little patch outside of town. Fortunately, today more and more people seem to be making the effort, for all kinds of reasons not necessarily relevant to chickens, to do just that.

Whether you are living dangerously on the pearl-gray side of the law, or soberly and legitimately in a rural setting, there will be no essential differences in your initial preparations for raising chickens for the flesh of their bodies. Obtained commercially, chicks are commonly sold as day-old, and at this age their most immediate requirement will be for warmth. Regardless of how many birds you start with, or whether you are rural or slightly urban, it is absolutely essential to provide a situation that can be maintained draft free at a controlled temperature, with sufficient roominess so that the chicks are not encouraged to bunch up. The number of chicks you start with will determine the size and arrangement of your physical facilities.

A big advantage the hobbyist starting with a small number of baby chicks has over large operations is that small size simplifies everything, from equipment, housing, and handling to problems of disease and vermin. *Small* is, of course, a relative term. When raising chicks for fryers, I would consider anything from half a dozen to about fifty as small. Over one hundred, unless you have a large family or a community project, I think you can safely say that the operation is beginning to get a little large, although still far from commercial in size. The initial investment in the cost of the chicks is not a major investment for the small operation. While the price per chick is reduced when purchasing in quantity, the savings is relatively insignificant when buying less than one hundred. The first consideration must be how many you and your facilities can handle; the second, of course, is how many you actually need or want.

Baby chicks are very small and take up little space. After all, a day-old chick is exactly the size of an egg, only a little less compact. However, they do not remain small for long, so do not be deceived by the space requirements of day-old chicks. On the most simple level, beginning with a small number, it is quite possible to keep them in a large box inside one's house for a week or two. One spring we kept a couple of inherited goslings inside the house in a large box during a spell of unexpectedly cold weather. A visitor (sometimes it seems we have few visitors) finally brought herself to ask why we had baby geese in our living room. With perfect composure my wife responded, "Because we have baby rabbits in the *kitchen*." While it is possible to nurse the hatchlings over the first couple of tender weeks inside your house, you must be a tolerant person and love little living things, or be passionate over fried chicken. Cleanliness is of the essence, of course, for the sake of all concerned.

More practically and permanently, a garage, or a chicken house if you should be so fortunate, provide the soundest basis for setting up a brooder operation. Sometimes facilities which are quite borderline may be made serviceable. One year, lacking facilities with the remotest potential for heating, I got from an appliance store the box in which a refrigerator had been shipped. It provided a surprisingly sturdy and roomy small-scale shed, which I put in an otherwise too drafty lean-to. Hanging over the middle of the box, I contrived a hover made of screen cut in a circular pattern and faced with aluminum foil. In the center of this hover I installed an infrared lamp for heat. Into this I introduced fifty chicks. The floor was lined with newspaper, which I changed daily, adding wood shavings on top of it after about a week. The corners were stuffed with wadded newspaper to discourage the chicks from crowding and wedging into them. After two weeks I cut a door in one end of the box and allowed them to run on the floor of the lean-to. They continued to

The Wholesome Chicken

retreat to the box for warmth, but after two weeks all the feed and water was on the outside. It was a make shift arrangement, and it would never be recommended by a professor of poultry husbandry, but it worked fine. When it was over, I composted what was left of the box!

Whether the facilities are a shed, a garage, or a chicken house, for the first few weeks it will be necessary to provide the chicks with some sort of brooder area, an area which supplies warmth and substitutes for the body warmth of the mother hen. There are two basic systems for supplying warmth: the entire building may be heated, or heat may be supplied by the brooder itself. The brooder

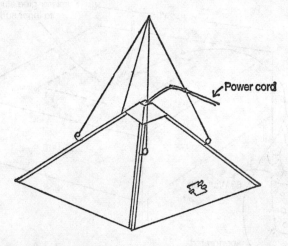

A typical form of commercially made metal brooder (hover). This one is suspended from above. Some stand on adjustable metal legs. Note the small door on the right for a thermometer. Most are heated with an infrared bulb hung in the center.

may be homemade, or it may be a commercial model. Small brooders are not expensive and are available through the large mail-order houses, or through farm-supply stores and feed stores. Although both do the job, commercially built brooders tend to be lighter and probably easier to clean than homemade ones.

If the house itself is to be heated, probably the ingenuity of the particular grower will decide how the heat is to be supplied. The advantage of heating the whole house is that it eliminates the need of brooders and makes it possible to use a larger part of the floor area for starting chicks. The disadvantage of heating the whole building

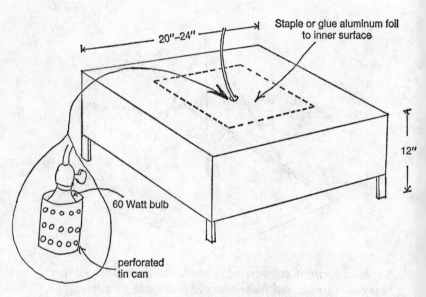

Cheapy brooder made from cardboard box fitted with wooden legs, heated with light bulb.

The Wholesome Chicken

is that it is apt to be more costly for energy, and it may be hard to maintain a uniform distribution of heat. For efficient heating, the building should not only be free of obvious drafts, it probably should be insulated as well. Whatever source of heat is used, care should be taken to eliminate danger of fire. If electricity is not available, the catalytic heaters such as are used by campers to heat tents should be safe. Probably less safe are the various kinds of kerosene heaters, although in days past (when chicken-house fires were very common) they were widely used. Brooders also used to come supplied with a kerosene heater, and even kerosene lanterns were used as a source of heat. If the whole building is to be heated, it is very important to protect its corners to prevent bunching. Curved or compacted pieces of corrugated cardboard serve very well. If cardboard is used, the material should be removed and replaced with clean every time a new batch of chicks are brooded. Even if the whole building is heated, sometimes brooders or low-hung heat lamps are used to encourage the chicks to stay away from walls and corners.

Before bringing in the baby chicks, the building or area being used should be thoroughly cleaned. If it has ever been used before for poultry, it should be disinfected. The floor should be covered with absorbent (but cheap) litter: wood shavings, ground corncobs, peat moss (sphagnum), etc. The choice generally is determined by which is locally available and least expensive. A floor which is not covered with litter tends to be colder and is more difficult to clean. If it is not intended that the room itself be heated, brooders, or a brooder system, must be provided for. If a brooder is purchased, it will include the instructions for operating that particular model. When day-old chicks are ordered, the temperature of the brooder (or brooder house if the whole space is to be heated) should be established at 90 degrees. Most hobby growers will find it more convenient and economical to start baby chicks only in the spring and summer months. If it is decided to start the chicks during cold weather, before bringing in the chicks the temperature should be established at 95 degrees. Thereafter, the normal procedure is to reduce the brooding temperature by five degrees a week until the temperature of 75 degrees is reached. If a brooder is being used, measure the temperature about three inches off the floor and three inches inside of the hover. Most brooders have a port along the outside edge for a thermometer; usually the thermometer is left permanently in place and is easily read without removing it.

If hover brooders, either homemade or commercial, are used, it is still recommended to maintain the temperature of the house or the brooder space at about 65 degrees. If the outer temperature is too cool, the baby chicks will tend to cluster under the brooder and not eat or drink properly. It is also possible that they will huddle together so closely that some may smother. If the temperature is too hot, they will be seen to hunker up and pant. The air in brooder houses may become dank and stale, although, for the hobbyist, ventilation does not tend to create severe problems for the small (i.e., twenty-five to fifty) flock. With larger flocks it is necessary to arrange for a circulation of air to reduce the humidity and maintain the oxygen, but without creating chilling drafts. In these larger areas, small exhaust fans may be installed. When brooding chicks in summer, it will conversely be necessary to protect them from temperatures in excess of 95 degrees. Except in unusual circumstances, fans which move the air, without directly putting a draft on the chicks, will be sufficient for cooling.

Equipment for feeding and watering may be either homemade or obtained from commercial sources. For their main feeding needs, a simple V-shaped wooden trough divided by a wooden dowel, to discourage the birds from getting their feet in the food, is quite adequate and is cheap and easy to make. The same system, translated into metal, is used in the galvanized-iron feeders purchased commercially.

There are many different varieties of waterers available. The problems presented in establishing a watering system for baby chicks is to provide them constantly with ample water without exposing enough surface for them to wet themselves, drown, or contaminate it. Waterers must also be easy to clean and convenient to fill, as well as hold an ample quantity for the number of chicks being brooded. If the watering facilities are inadequate, constant attention is necessary to make sure the chicks have enough water. For day-old baby chicks, the small plastic or metal caps that screw on the top of mason jars are quite adequate and very cheap. As the chicks grow, however, their requirements for water will soon make these small waterers unhandy and inadequate, and a second system for watering must be employed. Probably by their second, and at least by their fourth, week, the chicks will have matured sufficiently and their water requirement increased to the extent that a second system of watering is required. Therefore it is wise to have planned for and prepared these additional waterers at the same time the brooder is being put in order. Open pans, troughs, and the like are

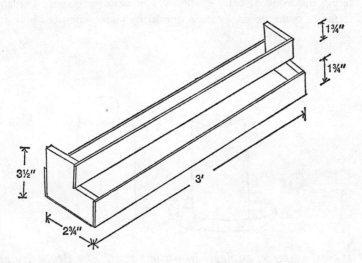

This simple homemade feeder can be used until the baby chicks are two or three weeks old. One of this size should be adequate for about fifty chicks.

not suitable. They quickly become contaminated, and chicks will be wetting themselves, if not actually drowning, in them. The best system to supply water in the quantity needed without contamination or wetting the chicks is some kind of gravity feed waterer. These come in a wide range of styles and sizes, from a gallon to fifteen gallons or so. Those designed to be hung from an overhead support are probably the cleanest and handiest, and result in least spillage. Their height will have to be periodically adjusted upward as the chicks grow. If suitable and proper feeding and watering equipment is obtained at the beginning, the second time a batch of baby chicks is raised, the only necessity will be to clean the feeding and watering equipment before the chicks arrive. About supplemental feeders with laborsaving qualities for use with larger chickens, more later. Such physical and mechanical problems as space, systems, and equipment should have been examined and resolved before actually ordering the stock of baby chicks. When evaluating space available in terms of space needs, provide one half square foot per chick for

A typical waterer, available in various sizes. The sleeve (right) slips over the filled tank, pressing a trigger which allows the water to flow into the watering pan. Vacuum controls the water level.

the first two weeks. Up to ten weeks old, they should be provided with at least one square foot per bird. At this age they will be roughly "frying size." The larger birds raised for roasters and kept up to about twenty weeks should have two or three square feet per bird. Older than twenty weeks they require four or five square feet each. These space requirements should be considered minimal; there are many advantages to providing greater space, particularly as the chicks become larger. It is distinctly desirable to allow the birds outside range, or at least exposure to the outside, by the time they are four weeks old, certainly by six, providing that the weather is suitable. It is this fresh air and exercise that many people feel are essential in making the home-grown chicken so distinctly superior to supermarket chicken. It is possible to raise the birds entirely in confinement; most commercially produced fryers have never seen the light of day. Home-sized "broiler factories" may be purchased at a modest price by mail order or through local supply houses. With such factories you can mature an appreciable number of

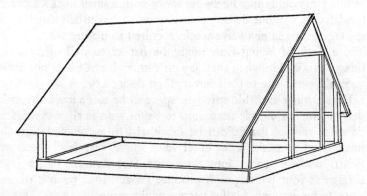

A simple frame and plywood range shelter 5′×7′ will protect fifty young chickens. If confinement is desirable, the frame may be covered with chicken wire.

fryers within very limited space—a garage, or even a basement, and on a year-round basis. If your interest is in quality rather than streamlined efficiency, this system will probably have little appeal.

If you are making do with existing facilities, the above figures will give a general idea of what the space you have available will allow in terms of number of birds. It is the space ultimately which will determine the outside limit on the number you can raise. Nevertheless, even if adequate space is available, most likely the hobbyist will begin with no more than twenty-five day-old chicks his first time around. Although raising fifty is little more trouble than raising twenty-five, the beginner will feel more comfortable starting with a smaller number. If you are going to build facilities especially for brooding chicks, keep in mind that eventually you may want to increase the number of chickens above whatever basic minimum you began with. Although you may only begin with twenty-five, keep in mind that at some time you will most likely want to raise more. In addition to enlarging the production of fryers or roasters, perhaps you may want to consider someday expanding your operation to include raising layers if you have a convenient rural location. While you could quite likely get away with a small flock of contraband fryers without detection in a more or less urban situation for as long as the ten or twelve weeks required to mature them, phasing into a flock of laying hens might be harder to pull off. For one thing, hens cackle when they lay an egg, and while it is not a disagreeable sound (quite the contrary!), it does carry.

Determining available space, or space to be made available, makes determining an outside maximum to begin with fairly easy. A fractional section of garage can be fabricated to accommodate such a small number as six or eight (I have raised that few fryers at a time), without going into wire-cage factories. Six large fryers dressed at four pounds each produce twenty-four pounds of meat —not to be despised. Half a garage might easily give you the minimum of ten square feet required to accommodate twenty chicks. Using your entire garage could yield you—but that way madness lies.

The Wholesome Chicken

When you have studied the space available, and you know what potential you have to work with, the next decision will probably be what breed of chicken to start with. For fryers or roasters, there are a number of heavy meat birds available to consider, including such older and familiar breeds as Rhode Island Reds, Plymouth Rocks, White Rocks, etc. All of those breeds produce a large and meaty bird, good flavored and efficient in terms of feed conversion. Increasingly, however, both commercial and amateur growers are raising varieties of crossbreeds, mostly crosses with a Cornish sire and a white bird, often a White Rock. Very popular is the Hubbard, a white heavy chicken, easy to grow, fast maturing, and an exceptionally efficient "converter of feed." Most of the meat chickens, by the way, particularly the crosses, are not as good for layers as the smaller breeds developed particularly for layers. It is not a good plan to carry over pullets of the crosses to form a laying flock. While they lay large eggs, the Hubbards, for example, are perhaps half as productive egg producers as White Leghorns, at the same time consuming 15 per cent more food. Before deciding on a breed to raise for combinations of meat chickens and layers, it is always a good idea to consult your county extension agent to find what seem to be preferred or more successful producers in your area. However, don't be afraid to be impractical and choose a breed simply because you like its looks. Differences are hardly likely to be catastrophic.

There are a variety of places to buy baby chicks. They may be ordered in most areas through mail-order catalogues—even seed catalogues—or through specialty poultry catalogues. During the spring and early summer, in rural areas, feed stores usually stock baby chicks on a regular basis and do not require ordering in advance. Perhaps there will be hatcheries in your area, where it will be possible to special order even unusual or exotic breeds of chicks. Your county extension agent will know, or you may find them simply in the yellow pages of your phone book. Baby chicks are almost always sold with a sex designation: 95 per cent cockerels, 95 per cent pullets, or "as hatched," in which case they will be almost evenly divided between male and female. It will depend on the breed which is least expensive, but for meat birds generally, the as-hatched chicks are most economical. The cost of baby chicks has varied so much in recent years that it is pointless to offer any guidelines to their specific costs. Certain breeds may command twice or more the price of others. Bought in quantity, baby chicks are cheaper than when bought in small lots. The first quantity unit which brings a reduction in price is usually fifty, so most hobbyists will not find the saving sufficient to be worth consideration, unless several growers get together to buy their chicks at the same time.

The largest single investment in the chickens which you raise for meat or eggs is going to be the cost of the feed. Unless your situation gives you unusual access to cheap sources of feed, most of what your chickens consume will be obtained from commercial sources. You may consult Department of Agriculture bulletins for specific nutritional requirements, but the commercial preparation may be generally depended upon to provide a balance of the required nutriments for specific periods of growth. If you decide to mix your own feeds, obtaining and grinding various grains, etc., Department of Agriculture publications give nutritional information on various raw feeds and guidelines for mixing proportions.

When your chicks arrive, you should arrange to have on hand a standard formula of starter mash. That's what you ask for: *starter mash*. It will contain 20 to 24 per cent protein. During the period of superrapid growth of their first weeks, chicks have an extremely high requirement for protein. For the first two or three days, it will appear that they are not consuming an appreciable quantity of feed at all, but by the end of their first week you will notice the increasing consumption begins to approach a geometric progression.

One of the most effective ways to feed the chicks for the first few days is to place the starter mash in a shallow cardboard container, such as the lids of the boxes the chicks were delivered in. Although this method of feeding is inefficient, in the sense that it is somewhat wasteful, the encouragement it gives to the chicks in stimulating them to start feeding more than outweighs the disadvantages. Very quickly—within a week or ten days at the most—it will be desirable to change over to a more efficient feeder of suitable size, as described above. While the starter mash will contain the grit necessary for the chicks to digest their food (I have heard farmers growl, "Do you realize you're paying for rock?"), it is a good idea to provide from the beginning a feeding station to supply them with additional grit. This grit consists of coarsely ground granite and is available at feed stores in several grades of coarseness. You will begin, naturally enough, with chick-sized grit.

Plan on about twenty pounds of starter mash per fifty chicks for the first two weeks. These same fifty chicks will consume another 250 to 300 pounds of starter mash in the next *four* weeks. The quantity of feed they put away as they begin to grow almost defies belief. Even before you actually price the mash, it is readily apparent that feeding chicks is not going to be cheap, and you will understand why commercial growers market their chicks as young as possible. Grit, fortunately, is very cheap: they will consume about twenty-five cents' worth of it in that time.

Unless you raise appreciably more chicks than the usual amateur, or engage in co-operative buying with other hobbyists, you will most likely not buy your feed in bulk, but by the sack. Different brands favor different sizes of sacks, but it is most economical to buy all feed in the largest-sized sacks available. Feed bought in bulk is cheaper, but the savings is hardly worthwhile unless you have convenient and adequate storage space. It is very easy for bulk feed to be wasted or contaminated, especially by birds or rodents.

Instead of starter mash, it is possible to feed the chicks partially or exclusively on a preparation known as "chick scratch." This is a closely ground mixture of various whole grains, mostly corn. It is much cheaper than starter mash, but contains a lower per cent of protein and is without the nutritional additives of starter mash. Chicks fed on scratch will show a slower rate of growth and development than those fed on starter mash, and probably you will notice a higher death rate. However, some people prefer to use the scratch because of the lower cost, or some because they feel that birds fed on grains exclusively, for their entire term, are better flavored or more nutritional. If you feed them scratch, it is absolutely essential that they be provided at all times with ample grit.

Chickens being raised for fryers should be switched at six weeks to a "finishing mash." This feed is also called a "broiler mash" or "broiler ration." The protein content of this mash is somewhat less than that of starter mash, and it will also be somewhat less expensive to buy. Continue to provide a feeding station, offering them chick-sized grit throughout the feeding period. Fed on standard starter and finishing mashes, at eight to twelve weeks (I prefer them to be no younger than ten weeks) you can consider them to be fryers and work your will on them.

The diet of chickens which are to be raised to a more mature eating size, the size usually designated as "roasting chickens," will vary from that fed the fryers when they reach six weeks of age. The different feeding schedule for those mature birds will be partly in the interest of economy and partly the flavor of the meat. At six weeks of age, an additional feeding facility should be installed, in which cracked corn will be offered to supplement the broiler mash. The cost of cracked corn is considerably less than that of the commercially prepared mashes, but since their requirement for protein is reduced by this age, cracked corn fed along with supplements is nutritionally adequate. Gradually increase the amount of cracked corn, proportionately reducing the mash, until at about twelve weeks of age you are feeding approximately equal amounts. It is possible to completely replace the mash with corn, but the time required to complete their growth may be increased by as much as one third. When feeding cracked corn in any quantity it is absolutely essential that the birds be given all the grit they can consume. To advance a rough notion of the total investment in feed for a finished bird, fryers—eight to ten weeks—will consume *at least* two and a half pounds of feed per pound of meat produced. Roasters will be kept until they are from three to five months old; at this time the large meat-producing birds should weigh six or seven pounds, and the crosses as much as eight to ten pounds. The

The Wholesome Chicken

longer you keep them, the heavier the proportion of feed to meat will be. It becomes quite clear why roasting chickens sold commercially command a price so much higher than fryers and why very small fryers are frequently priced relatively low.

It is probably not something one would care to try the first time around with raising chickens for meat, but at some time the advantages of producing capons should be considered. A capon is a castrated cockerel. Even though the first notion that crosses one's mind is "Where in the heck do you start?" it is not a difficult procedure, and the instruments are inexpensive and uncomplicated. The operation begins by making a small incision on either side of the rib cage. The ribs are spread slightly and held apart with a clamp. Using a device somewhat similar in appearance to an old-fashioned button hook, the testes are snared and removed. Simple caponizing kits are available through farm and poultry catalogues or at poultry and feed stores. The kits include specific instructions, but it would be wise and help the self-confidence to consult a county extension agent who might even be willing to supervise your first attempts. There is a Department of Agriculture publication on caponizing which you can probably obtain from him. You may also find someone locally who does the operation and learn from him.

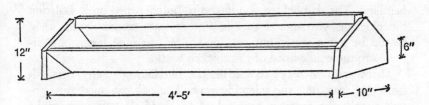

This type of wooden feeder is inexpensive and easy to build. The size can be varied according to need. The same design on a smaller scale can be used to feed baby chicks.

The operation is usually performed when the cockerel is three to five weeks old. If the chicks are older than five weeks, the advantages are mostly lost. After allowing a couple of days in isolation for recovery, caponized birds require no special attention. They will grow slightly larger and distinctly plumper than uncaponized birds, and the finished birds will be more tender, some contend more flavorful, than other chickens. They will also retain their tenderness for a longer period, up to as long as a year. It seems that capons are not as efficient in food conversion as uncaponized chickens, and will probably cost slightly more to "finish" than other chickens. Commercially, capons command a premium price and are not usually easy to find. If you are interested in raising roasting chickens, capons are certainly a possibility worth considering.

When the time comes for "harvesting" the crop, what is no doubt the least pleasurable labor of raising chickens for meat becomes necessary, the butchering. If you prefer, in most areas a butcher shop or commercial poultry operation can be found which will, for a small charge, turn live birds into dressed birds, ready to eat, freeze, share, or whatever. If you really want to get involved with the production of your own food, however, most likely you will want to do it yourself. You will find, after you have done it once or twice, that it's really no big deal.

Probably you will want to kill your fryers according to size so most likely will not butcher them all at once. First, then, you will select the largest third of your flock, or however many you feel like handling. The day before you intend to butcher, pen up in isolation the chickens you intend to kill. For at least twelve hours, they should be given water but not food. Alas, no final banquet is allowed to the condemned of the chicken world.

There are two ways to initiate the butchering cycle: the old-fashioned way, and the scientific way. The old-fashioned way entails simply the outright removal of the head. If you wish to be completely old-fashioned, you may "wring the neck": grab the bird by the neck, and give the body a couple of rapid and determined swings. As Mark Twain once put it, hoist it by the neck, very quickly! The body detaches quite easily. Less primitive, but still direct and certainly traditional, the head and neck may be stretched on a chopping block and the neck severed with an ax, hatchet, etc. Being traditional but not primitive, I use an army-surplus machete; with the long blade, the headsman's stroke requires little accuracy. As everyone knows, when the head is chopped from a chicken reflex action will cause the bird to flop vigorously for as much as a minute. To restrain this action and prevent possible bruising of the bird, I always thrust the body into a large bucket and restrain him tightly. To make this method work, you must have a firm grip on the ends of the wing feathers as well as the legs. Their muscular strength is considerable, and if a wing or wings get away from you, it becomes very messy.

The scientific way of killing a chicken, and from all practical standpoints undoubtedly the best, begins by hanging the bird up by its feet (by a stout string attached to a nail, etc.). Several may be attended to at one time. Instead of an ax or hatchet, a knife with a long, slender, very sharp blade is required. First, make a deep cut in the neck just below the head, severing the veins but not the neck bone. Then insert the knife in the mouth and pierce deeply the roof of the mouth far to the back, and rotate the blade. This technique has two advantages. Hanging upside down by the legs allows the blood to drain quickly and completely from the body. Second, piercing the brain through the back of the mouth causes the feathers to loosen immediately. If one works with relative speed, it is possible to dry-pick—i.e., remove the feathers without the use of hot water—which is both neat and convenient and also leaves a skin that presents an attractive white appearance. Most beginners, however, will prefer to use the more primitive method on their first trial, and decapitate.

If you choose to cut the head off, it is wise to scald the chicken immediately after before attempting to pluck it. Otherwise plucking will probably be slow and disagreeable, and you'll be lucky to finish without tearing the skin badly. The water the chicken is to be scalded with must be neither too hot nor too cool. Too hot, and it will tend to "set" the feathers, making them hard to remove and probably discoloring the skin as well. Too cool, and the water doesn't do the job at all. An old-timer first explained to me the system which I always use to gauge the correct temperature. When the water begins to smoke, brush the surface quickly with the flat of your palm. Do it again immediately. If you don't want to do it a third time, the temperature is just right. Otherwise, a little more heating is required. If you don't want to do it a second time, the water is already too hot. While I never use a thermometer to test the water, if you wish to determine the temperature scientifically, the thermometer should give a reading of between 130 and 140 degrees.

After decapitation, as soon as the muscle spasms cease, immerse the chicken completely in the bucket of hot water, hold him under for about thirty seconds, gently pumping him up and down to allow the water to completely circulate under all the feathers. Test a wing feather; if it comes out easily, the bird is ready to pluck. If it resists firmly, dunk the bird again for another thirty seconds. With a good scald, as they say, the short feathers of the breast, back, and thighs should almost roll off. Lay the chicken on a convenient surface, or hang him by his legs, and, beginning with the coarse feathers of first the wings and then the tail, pluck. The pinfeathers, if it has any, can usually be removed most conveniently after all the rest of the feathers have been plucked. I usually do the final "pinning" inside at a sink, just before dressing. The edge of a not-too-sharp knife is a useful tool to assist in this operation. You will most likely notice a scattering of small hairs over the body. After all of the pinfeathers have been removed, these can be singed off over a gas burner, or with a tuft of cotton soaked in alcohol. Another common way is to crumple a page of newspaper, set it on fire, then run the body of the bird through the flames very rapidly a time or two, rotating it at the same time. This latter technique may cause a slight discoloration, especially if done too slowly. And, obviously, whatever technique is used, avoid creating a fire hazard.

The dressing is best done at a sink. With a sharp knife, make an

The Wholesome Chicken

incision at the base of the neck and work the crop loose and out, then cut free. Turn the body around and make a horizontal cut halfway between the keelbone (the breastbone) and the vent (or anus). Cut completely through the skin, but move carefully to avoid cutting the intestines. Carefully cut around the vent, then work all of the viscera out. This will include the liver and gizzard, but you will have to reach inside to remove the heart and "stones" (testes, if it is a cockerel), which are attached, but not firmly, on either side of the middle of the back. The lungs are the most difficult to remove since they are located in two recesses far forward and covered by a tough membrane. It is easiest accomplished by breaking the back downward just to the rear of the ribs to enlarge the access, unless the chicken is a large roaster. With the tip of the knife, cut the membrane covering the lungs, and with the point of the blade, work the rear lobe free. A finger can be slipped easily under the lung, and the lung worked free. When the lungs have been removed, find the end of the windpipe and, tugging determinedly, pull it out. If any remains in the neck, remove it from the front. Where the tail—the "deacon's nose"—joins the back, you will locate the oil gland, which all birds have, and which secretes the oils with which they groom and condition their feathers. Make a deep straight cut to the bone on the forward edge of the gland, then a slanting cut under the gland from the rear. Discard. Also remove and discard the feet.

Attend to the giblets last. To prepare the heart, with a straight cut remove the top veins (i.e., about the upper one fifth of the organ). To prepare the liver, find the gall bladder and remove it gingerly. The fluid from a broken gall bladder imparts a vile and bitter flavor to anything it touches. It is best simply to cut away the small portion of the liver to which it is attached. Open the gizzard by making a long cut two thirds of the way around the thick, muscular rim. The lining, a tough, semitranslucent membrane, is worked loose from the muscle and discarded. Rinse all the giblets carefully, taking particular care to rinse clean all of the inner surface of the gizzard.

Backyard Poultry Raising

The drawn body of the bird should now be rinsed thoroughly in chilled water. It is probably my own private eccentricity, but I prefer to allow a chicken to remain overnight in the refrigerator, soaking in cold water to which a bit of salt has been added. If it is a frying chicken, I always cut it into frying pieces before soaking it. There are instructions for freezing chicken in various handbooks, manuals, and government publications on home storage of foods. I freeze fryers, cut up, in half-gallon milk cartons (properly cleaned, of course). When the meat has been packed inside, I fill the carton about four-fifths full of water, recrease the top, and secure the seam with three or four staples. Not everyone recommends freezing the meat in water, but for fryers I prefer it. Roasting chickens I simply wrap in two layers of freezer paper, tape, and store. If you wish to freeze the giblets, put them in small wax bags or sandwich bags and place in the carton or inside the body of roasting birds.

It is a rather curious fact that while butchering fowls is a chore no one prefers to do, the final preparation of the bird at table is a task of honor, and the carving is an act of ceremony. Along with the "nouns of assemblage" for plural numbers of animals our ancestors troubled themselves to contrive and memorize ("a gaggle of geese," a "crowder of cats," a "kindling of rabbits") came, for ceremonial purposes, the "terms of address" to a fowl. So the carver at table would "lift a swan"; "rear a goose"; "dismember a heron"; "unbrace a mallard"; "wing a partridge"; "display a quail"; "unjoint a bittern"; "unlatch a curlew"; "break an egret"; and "thigh a woodcock." Of no relevance whatever to the topic of fowls, there existed a parallel system of "terms of address" to fish. These include the indispensable "barb a lobster"; "chine a salmon"; "string a lamprey"; "culpon a trout"; "scull a tench"; "transon an eel"; "side a haddock"; "splay a bream"; and, perhaps most instructive of all, "tame a crab."

These terms contributed to ennobling the final preparation of fish and fowl, but the butchering necessary before it goes to the kitchen retains a status that is humble and perhaps even humbling. However, it is not a difficult job, and need not be particularly unpleasant.

The Wholesome Chicken

I first undertook the chore when a quite small boy and have taken the whole matter in stride all of my life. My young daughters now help me when butchering is to be done, and if you care to consider the educational value, they have learned a great deal of practical anatomy and, more important, the sense of how a living organism works. Plus, I believe, a genuine awe and respect for what goes into making the three dimensions of a living thing.

The Egg and How to Get It

It is quite possible to combine raising chickens for meat with raising chickens for eggs. In fact, from time out of mind, until the last forty years or so, chickens for eating were largely a byproduct of egg raising and more or less exclusively a farm and farmyard production. The old-fashioned diversified farm was a remarkably efficient little ecosystem. In the barnyard, for example, whole grain was fed to a small herd of fattening cattle (and to the farm horses). It was inefficient feed, and consequently a quantity of it passed through the animal undigested. A couple of hogs might be kept in the farm lot, who sought out and browsed off this waste grain (plus other rations, mostly scrap, occasionally given them), and grew up to provide their own contribution to the family economy. Some grain, etc., passed through the pigs. The farm's flock of chickens browsed on what fell through plus the insects the whole operation generated and, during a large part of the year, not only survived but thrived pretty much on what they could scrounge and forage. In the spring, the combined excreta, which even the sparrows (which had contributed their own bit) could no longer extract nourishment from, was loaded into the manure hearse and spread on the fields. Which, in turn, responded with a crop of corn even better than the year before.

In the turning cycles, we have reached the point where manure is an undesirable waste product, a pollutant no one seems to know what to do about. In days not too far in the past, I have heard farmers, with an almost moral indignation, condemn people who kept no livestock. Such farmers were robbing the soil, they said, and returning nothing to it. One such farmer, I remember, provided an apocalyptic vision of a farm turned into a gigantic pit, bounded by its fence lines, by an unthrifty heir who subjected it to such abuse and depletion.

The days of such neatly closed ecocycles are not likely to return, and probably only in the rosy vision of nostalgia could it be conceived desirable that they should return. But, given even modestly adequate circumstances, raising a few chickens for eggs gives an opportunity to restore a segment of such a life system for one's private experience and provides pleasures that extend beyond the commodity itself. Hens kept for eggs represent a kind of animal permanence and continuity that are different from the more ephemeral experiences with chickens raised for meat. Of course, given sufficient space and adequate facilities, there is no reason why chickens cannot still be raised for both meat and eggs, even, in some situations, restoring the meat chickens to a kind of byproduct. The facilities for raising each are different, but certainly not incompatibly different.

There may likely be a difference between the breeds selected for each purpose. If you wish to combine meat production with egg production (as distinct from raising them as parallel but distinct operations), the range of choice in selecting a breed is immediately narrowed. While pullets of the large modern crossbred meat producers, such as the Hubbard, will provide fine eggs, they are inferior layers and not at all efficient in terms of outlay for feed. The difficulty is simply that to produce a small number of eggs you have to sustain a very large hen. The combination chickens, the old standard varieties, still provide an acceptable compromise if you wish to combine meat production and egg production with a single strain. Such breeds as the Rhode Island Red, Plymouth Rock, White Rock, Buff Orpington, etc., are quite acceptable egg producers for the back-lot operation and provide a nice large-bodied meat bird. Also they are all chickens that are attractive to have around. Most of these breeds produce a dark egg. In egg production they are not as prolific as those hens bred exclusively as laying hens, but for one more interested in quality to provide for his own family consumption, they give a grower with a small flock no reason for complaint. Raised for meat, they will be found to be not as efficient in food conversion and rate of growth as the modern breeds developed particularly for meat production. But they are plenty satisfactory raised in a small unit for home consumption. I have known growers who believed their relatively slower rate of growth produced a better-flavored chicken.

Not too long ago dark-shelled eggs were "docked" for their color, and only pure-white eggs commanded premium prices. People who have a decided preference now pay premium prices for dark-shelled eggs. In England, a light-shelled egg is virtually unsalable. If there's a difference, honestly I can't tell it, but, for whatever irrational and illogical reasons, I prefer dark-shelled eggs. However, probably the most efficient egg-laying chicken ever developed is the White Leghorn, and she lays pure-white eggs. If the market at large preferred dark-shelled eggs, no doubt a hen laying dark-shelled eggs with equal efficiency would be developed.

The Leghorn is a small-bodied bird, slender, with a sort of narrow pigeoned breast. The meat is a little stringy and not especially good flavored, what there is of it. You would hardly consider raising her for meat, even as a compromise. But she lays big white eggs, and she lays lots of them. Having a small body, she uses a smaller amount of feed in sustaining herself than a larger laying hen, and consequently a higher per cent of the food she consumes is turned into eggs. Recently an even smaller-bodied Leghorn has been developed, laying the same sized eggs but even more efficiently than the regular breed.

Leghorns have a curious disposition. They are unusually nervous and, in a quite unmetaphoric sense, flighty. Even if the birds are virtually hand-raised, they seem never really to get used to people. When you walk into the chicken house or chicken run, they tend to scatter like quail. Also, they are quite accomplished fliers. A fence which is perfectly adequate for other breeds of chickens will be easily over-flown by a frightened, or simply determined, Leghorn. They also have a tendency to startle when you walk into a chicken house where they are kept in open confinement. A quite workable expedient is simply to knock on the door loudly a couple of times and pause briefly before going inside. Although it makes one feel a trifle foolish and Victorian to knock on the hen-house door before entering, as if some of them might otherwise be exposed in dishabille, it helps teach them not to have hysterics just because you come into the chicken house.

If you wish to establish a flock of laying hens, there are three, or really four, ways to begin. As with chickens being raised for meat, you can begin with day-old chicks. These may either be sexed or unsexed. Sexed (95 per cent pullets) will cost about twice as much as unsexed or "hatchery run," which will turn out about fifty-fifty males and females. With Leghorns it is a clear advantage to buy sexed chicks, because the cockerels have so little to recommend them for eating purposes. The second method of establishing a laying flock is to buy "started" pullets. These are usually between six to eight weeks old, old enough to no longer require brooding. These pullets are, of course, closer to laying age and have in addition survived the period of highest mortality rate. Their cost is considerably higher than day-old chicks because there is already a considerable expense of feed and care invested in them. If you intend to raise nothing but laying hens, to buy pullets already started will eliminate the necessity of owning brooding facilities, brooding equipment, and special-sized feeders and waterers. Under such circumstances, it is clearly an advantage to begin with started pullets.

The third way to obtain your laying flock is to buy pullets which are ready to lay, which are sold at sixteen to twenty weeks, or sometimes slightly older. They can be expected to commence laying almost as soon as they get settled into their new surroundings. Ready-to-lay pullets are most expensive of all in initial price. Again, buying them at this age has the advantage of reducing requirements in equipment and facilities as well as care and natural shrinkage. They are also ready to be fed laying mash, which is less expensive than starting mash. Taking all considerations into account, this is perhaps the most feasible way to begin a small flock. But you do miss out on the fun of raising the baby chicks.

The fourth way to get started is to buy mature laying hens. Sometimes you will find them advertised in the classified ads of newspapers, sometimes they may be obtained at farm auctions, or sometimes simply ask someone with chickens if they will sell you a few. If there is a commercial egg operation conveniently located, you will probably find they have some they will sell you. Hens which outlive their efficiency are usually sold in wholesale lots to canneries, particularly to soupmakers, but often a few can be purchased by an individual, and very cheaply. Remember, however, that the reason these hens are being culled out of the flock and sold is that they are no longer efficient egg producers. Nevertheless, a small-scale hobbyist may well have more liberal standards of efficiency and be able to overlook the relative inefficiency in favor of initial cheapness and convenience of procurement. Particularly with a White Leghorn, it is fairly easy to judge the laying potential a hen may have remaining. These techniques are discussed below, or take with you someone who has had experience with laying hens and knows how to judge them.

If you begin with day-old chicks, the feeding schedule and type of care required for pullets being raised for layers differs somewhat from the methods employed in raising chickens for meat. Initially, the day-old chicks are handled in exactly the same way as potential fryers. At six weeks, however, those intended to become layers are changed to a finishing or growing ration, and, at the same time, have added to their diet grain—cracked corn, wheat, millet, or combinations. What is locally available and least expensive will largely determine the choice. Begin by feeding (in separate feeders) a hundred parts of mash to ten parts of grain. Gradually increase the grain proportion until they are getting equal parts of each. When

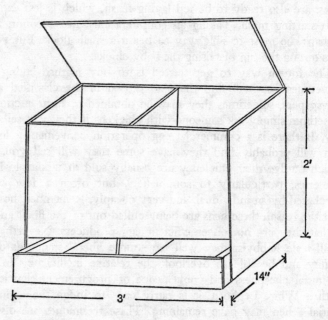

This two-compartment hopper with a hinged lid serves grit on one side and oystershell on the other.

the pullets reach the age of eighteen to twenty weeks, gradually reduce the growing mash and replace it with laying mash until, by the time they are twenty-two weeks old, the growing mash has been completely replaced by the laying mash. As they begin to lay, continue to feed the same proportion of grain and laying mash. Ideally, most hobby ranchers will arrange their chicken facilities to provide for a certain amount of range feeding or foraging. This will reduce the intake of purchased feed and most likely improve the health of the birds and enhance the flavor and appearance of the eggs.

As when feeding grain to roasting chickens, the pullets must be provided with a constant supply of grit. By the time they are twenty-two weeks old, they should be given the coarsest grade of chicken grit. It is a good idea to buy, or to make, a feeder or hopper which will make the grit readily available at all times. Depending on the size of your flock, this feeder will only need to be refilled once a month or so. In addition to the feed and grit, laying hens require a great deal of calcium for good-quality eggshells. Eggshells are almost pure calcium and, as much as possible, they should be recycled. I have heard it argued that allowing the hens to eat eggshells tends to introduce them to breaking and eating eggs in the nests, but this is something that has never been troublesome to me. The problem is more likely to occur under conditions when hens are kept in close confinement. At any rate, the hens will need a supply of calcium supplement. The standard source of this mineral is crushed oystershell or crushed limestone. Like the grit, it is bought at the feed stores and is happily not very expensive. Also, like the grit, it should be made freely available at all times.

The word "small," as in "a small flock," is certainly a relative term. Adequate size is determined by a factor and a function. The *factor* is how many eggs you feel you need, and the *function* consists of the capacity of your facilities and your feeding plans. If you expect your hens to lay efficiently in the winter, or even to lay at all, they must be confined to a heated facility in which you can also supply artificial light. Maximum efficiency comes from maintaining artificial lighting during a good part of the year. Light stimulates the hens' pituitary glands and encourages laying. For year-round egg production, they need fourteen hours of light per day. An automatic timer may be installed to turn lights off and on. Or, a low-watt (i.e., about 25 watt) bulb may be hung with a reflector in a small chicken house and left on constantly, beginning about October and ending in April. If you have a chicken house, but provide no heat or artificial light during the winter months, very few eggs will be layed. There is nothing morally bad or personally reprehensible about this practice; innumerable farm flocks are tended in exactly this way. It must be kept in mind, however, that the hens are eating about the same amount of food during the winter whether they are laying or not.

A family-sized flock consisting of half a dozen Leghorns may pro-

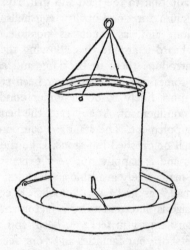

This metal overhead feeder is suspended from the chicken house roof, some eight to ten inches from the floor. Feed, poured in the top, gradually runs down as the chickens eat from the pan. A single filling will last a small flock several days.

vide perhaps two or three dozen eggs per week during the warm months. A commercial producer would expect more. In my experience, after a while the family egg consumption begins to taper off. If you choose, you can sell or trade the surplus or simply give them to friends. I find fresh farm eggs a particularly pleasant sort of thing to give to people, just as I enjoy giving fresh garden produce. Like bread cast on the water, often these gifts return fortyfold.

On a completely personal level, I retain a childish pleasure in the physical act of gathering the eggs. It's a pleasure I allow my family to share, but seldom others. Consequently, for that reason alone I find I need more laying hens than we need eggs. Six hens are more than adequate to provide for our kitchen needs, but with only that many hens, on any given day I find that I simply am not through gathering eggs; a dozen hens stagger us with the quantity of their eggs, but the pure pleasure of gathering them is worth the extravagance of the extra hens to me.

A smallish flock of six to a dozen hens gives the hobbyist many advantages over larger flocks. Many of these advantages are obvious —the general convenience of smallness, lessened problems with disease, reduced worry about efficiency in total cost accounting, and so on. But having only a small number to care for offers additional advantages in opportunities to supplement their diet with varieties of feed that will cost little or nothing. If you are a gardener, unending selections of garden scraps—overgrown lettuce, cabbage trimmings, corn husks, overripe vegetables and fruits of all sorts— provide a summer-long food supplement. In addition, weeds pulled from flower beds, prunings, and lawn clippings not only give supplemental nourishment for the birds but also provide a convenient system for disposal of these wastes. Also trimmings from the kitchen—varieties of scraps, bread crusts, and certainly eggshells— can make a considerable contribution for a small flock. There is also a good chance that making contact with the produce manager of a local store will give you full-season access to enough vegetable wastes to provide a very significant supplement to the diet of even a modestly large flock.

If you have a gardening spot, apart from garden waste it is quite possible to plant and harvest a major part of the food-stuff for a small flock. Unless you have a large area available, it is not very efficient utilization of space to attempt raising such grains as wheat, corn, etc. Sunflowers, however, can usually be grown in useful quantity, largely by making use of waste space, if you employ a little ingenuity. Because their space requirement is vertical rather than horizontal, they may be planted against walls and fences and use soil that would otherwise be accessible only to weeds and grass. They are also attractive in their own right. Try growing them around the fence of your chicken yard—planted just outside, of course. And try growing them in your garden for living bean poles—one sunflower seed, three climbing beans. This also saves the trouble of erecting bean poles and putting them away in the fall. The huge turniplike vegetable called a *mangle* is sometimes grown for chicken feed. They are so big that a sizable weight of them can be grown in a small space. They also keep fairly well. Just throw an occasional one in for your chickens to peck at. Calculate that a White Leghorn hen will eat about ninety pounds of feed a year, a heavier hen about 110–120 pounds per year. For a flock of six it may be quite possible to get by purchasing only a relatively small amount of feed if you make use of scraps, scrounge a little, and grow a little. The larger the flock, clearly the less feasible it becomes to turn waste into eggs or grow a little feed on the side. Whether it is from inherent miserliness or just a pleasure of independence, I find a great satisfaction in sustaining a laying flock on the food which I find the means of making available rather than feed which I simply buy.

Laying hens have a quirk of temperament, technicality of biological balance, or whatever, which makes them very sensitive to changes in their diet. Any sudden change of their feed is likely to result in a falling off or complete cessation of their laying. Consequently when shifting to any new or different food, the change needs to be made gradually. Adding greenery to their diet as it becomes available, however, does not seem to affect their laying adversely. When your pullets begin to approach laying age, eighteen to twenty weeks old, the main diet which you intend to supply them should have been established. If you must deviate from it later, anticipate the change so you may shift feed by degrees.

The first eggs your pullets lay will very likely show a few surprises. Sometimes their initial efforts will be puny little husks, occasionally no larger than marbles. The very small eggs sometimes lack a yolk or perhaps will have nothing but a yolk. Some will be perfectly normal, except for being about half normal size. Frequently the first eggs will be oddly shaped. I have seen some that were nearly cubes. Also the shells will frequently be rough and uneven, exceptionally thin, or perhaps with no shell at all. An occasional chicken will all her life lay nothing but eggs with wrinkled shells, and occasionally one will habitually lay shelless eggs. Not all pullets lay unusual eggs the first time around, but most will lay smallish eggs for their first week or two. On the other end of the scale, from time to time most any hen will just happen to lay an egg with a double yolk. Any extra-large eggs may be suspected of having more than one yolk, and I have seen eggs with three yolks. Commercial growers always sort out these eggs by candling, and, although common among laying hens, double yolks are virtually never found in the eggs you buy. Their oddity makes them a sort of treat to the home producer. They never fail to fascinate children.

Caring for eggs is about as simple as caring for potatoes if you discount their relative fragility. Like potatoes, eggs keep better if they are not washed, but since most hobbyists deal in a commodity that is by definition not intended to be kept long, this is not a rule that needs to be taken too seriously. I routinely wash unclean eggs. Like potatoes, eggs keep best stored in a cool, dark place. Unlike potatoes, eggs should be stored small end down, which is supposed to prevent the yolk from adhering to the shell and consequently rupturing when the shell is broken. One should not keep eggs around long enough for this to be likely to happen. If eggs are in great surplus, it is possible to freeze them (but not in the shell). Break the eggs and stir the white and yolk together (don't beat) and freeze them in appropriately sized containers. Yolks and whites can be separated and frozen individually, but some of the qualities of texture will be lost. Consult one of the various guides to home canning and freezing for additional tips and techniques.

Our rural grandparents commonly preserved eggs from the prolific summer for the meager winter in a solution called *water glass*, consisting of the chemical sodium silicate dissolved in water. A layer of the solution about two inches deep was poured into the bottom of a crock and allowed to become firm. The liquid soon sets up like gelatin. A row of eggs was placed, small end down, until the bottom was completely covered. Then water-glass solution was poured over, covering them with about two inches of fluid. As soon as the fluid set up, another layer of eggs was added, and another layer of water glass, and so on. Kept in a cool place, eggs cared for in this way would survive all winter. By spring, when an egg was removed from the water glass and broken, its liquid content would be found to have been reduced by a fifth or so, but in most respects they were dietarily satisfactory if not "farm fresh." Similarly, eggs were sometimes packed in crocks and covered by warm (not hot) lard. As the eggs were removed, the lard was salvaged for other uses, including frying the eggs.

The size and efficiency of a home laying flock will also influence or determine how long a laying hen is to be kept. It is possible to make this determination strictly on the basis of a time schedule, keeping them for a year, then replacing the whole flock at one time. The first year of their lives will be their most prolific period of egg laying and also their most efficient time in terms of feed conversion. Plan ahead in obtaining a replacement flock either by raising chicks, started pullets, or finished pullets, and make a complete switch as soon as the new pullets begin to lay. It is generally wise to buy a variety of different-colored chicken rings (ask for them by name at your feed store), which are slipped over the feet of young birds to keep exact track of when they were added.

Some time toward the end of their first year, hens commonly will molt. When a hen molts, or drops her old feathers and grows a new coat, she will stop laying. Sometimes after molting, a hen will lay as well as she did before, but usually their production will never again be that high. Sometimes a hen will never lay again after her molt. A flock that has been loosely maintained and stops laying during the winter can be tricked into molting in the early spring, which should put most of them back into production. Pen them up in their house for three days, curtaining all windows to keep out the light. Give them all the water they want, but no feed. On the fourth day, uncurtain the windows and give them light and begin feeding grain only, such as cracked corn, wheat, oats, barley, etc. In two weeks return to your normal feeding schedule. They should all molt, and as soon as their new feathers grow in they should begin laying again. If all goes well, they should begin laying about two weeks after the molt.

Whether or not you intend to keep your hens for a second year will to a great extent depend upon the individual birds. You well may keep some or most, but probably not all. It is important to study the hens and learn to distinguish the signs which indicate which hens are laying, and how well, and for how long, and which hens are nonlayers. Hens which do not lay are given the apt designation of *boarders;* they eat and sleep and enjoy the premises, but they make no contribution to their own upkeep. The process of identifying these boarders and removing them from the flock is called *culling*. Those removed are called *culls*.

White Leghorns are probably the easiest of all hens to cull because they have such prominent combs, and the condition of the comb is a prime indicator of their laying condition. The comb of a chicken is a secondary sex organ, and in a laying hen, as the ovaries are activated, the comb gets an additional supply of blood. Consequently, when a hen is laying, her comb will be full and large, and brilliantly red. Because of the size of a Leghorn's comb, these characteristics are so prominent as to be almost unmistakable at a glance. With other breeds the same changes in the comb take place, but since they naturally may have a comb that is smaller or less prominent, it may be inconclusive attempting to determine their laying condition on this criterion alone.

A laying hen will draw pigmentation from her skin for deposit in the eggs. Therefore the skin of a nonlaying hen tends to be yellow. The skin of a hen that is laying well and has been laying for some time will become bleached out and white. Likewise, the legs (shanks; the part that attaches to the feet) of a nonlaying hen will tend to be thick and the skin scaly, with a strong tinge of yellow color. The legs of a laying hen will be more slender, the underlying bone structure prominent, the skin a more smooth texture, and the color pallid in cast. The fading of the color tends to follow a pattern. The color will fade first around the vent and from there across the body, with the pigment of the legs fading last. Consequently, when the legs and shanks are light, usually it indicates that the bird has been laying for quite a while.

The vent and the pelvic-bone structure around the vent also offer good indication of the hen's laying status. The pelvis of a laying hen will spread to about twice the width of a nonlayer. The exact measure of the spread will depend on the breed and size of the bird. Measure the spread of the pelvis with the width of your fingers. After comparing a few birds of the same breed, you should find this to provide a quite accurate scale. The vent itself also gives additional indications of whether or not the hen is laying. The vent of a laying hen should be pale in cast, its appearance moist, and its size large. A nonlaying hen will have a small and puckered vent, usually yellowish in color, and with a dry, even scaly, appearance. Often the general appearance of a hen will give an indication of whether it should be culled, or at least be given closer examination. A hen will not molt while she is laying, so often the feathers of a properly industrious bird will become worn and bedraggled. The nonlayer may molt and betray its nonthriftiness by a fine and well-trimmed suit of new feathers. If you keep your layers over a second year, you should keep an eye on an individual you question on these grounds and give her a chance, however. She may be just getting a second wind after her molt.

The above features to watch for suggest a definite sequence of examination. A well-groomed hen with a comb that looks suspiciously shrunken, or a hen that displays only a shrunken comb, should be caught and given further examination. If other indications confirm the initial suspicion, the hen should be culled.

Since she has been lavishing the feed she has eaten on her own body rather than on egg production, a cull is likely to be quite fat. And a plump but unproductive hen provides a fine beginning for a variety of most excellent stewed-chicken dishes. Even a Leghorn, which is certainly not a first choice for an eating chicken, can produce a creditable swan song with dumplings. Kill and dress an ex-laying hen as you would any other chicken, but, remember, it will probably take a good long simmering before she comes out tender.

The Chicken at Large

As is the case with most domestic animals, no one knows for sure when chickens were first domesticated, or even which wild species were the ancestors of the domestic breeds. Perhaps they originated in eastern Asia, where wild fowls that closely resemble some of the older domestic breeds such as the game chicken still live in the jungles. Nor is there any way of knowing whether they were first domesticated for eggs or for meat or simply as pets. Wherever the ancestral stock originated, or why, by the time history picks up the record, chickens were widely distributed, no doubt far extending their original range. And at least since historical times they have been widely kept for both eggs and meat.

Backyard Poultry Raising

Certainly one of the early interests in chickens attended the fighting instinct of roosters. They are one of the few animals with both the equipment and the desire to fight a member of their own species to the death. The ancients appeared to have been particularly enthusiastic breeders of fighting chickens, as well as gamblers on the outcomes of the fights. In fact the ancients seem to have been enthusiastic fanciers of fighting anythings. Plutarch speaks of Mark Antony's exasperation because the fighting quail of Augustus always beat his own. Words which have entered the language from cockfighting attest to the interest our English-speaking ancestors had for the sport. Such words as *cockpit*, perhaps *plucky*, and probably *cocky* relate to the fighting place and to the temperament of the contestants. The expression "show a white feather" reflects a bird's loss of the keen dedication to fighting through inferior breeding. It is further probable that the first efforts of serious selective breeding grew out of an interest in improving the fighting strains. Only much later were the same principles of selection likely employed in building a meatier chicken or improving the egg production of hens.

While cockfighting is illegal in many countries, including the United States, it continues to be conducted in clandestine pits, particularly in rural areas. It is not a sport which I in any way condone, but the chickens bred to compete in it are creatures worth marveling at, and even fancying for their own sake. It is not farfetched to make a comparison between them and the breeds of bulldogs developed when the gory pastime of bullbaiting entertained the sporting set of the sixteenth and seventeenth centuries.

The game cock, or game chicken, can be considered essentially a strain of the bantam. Over the course of time I have owned several, purely for the pleasure of having them around. The most remarkable one I ever owned was given to me when I was a small boy. It found its way to me from the flock of a man who the law had forced out of the game-chicken business. Perhaps there is a basis in fact for the banal crack about the chicken inspector. The cock was nearly full grown when he came to me but had not been prepared for a fighter; he still had his comb and his full spurs, which grew to an astonishing length as he grew older. Eventually they became curved scimitars, slender but needle sharp, over an inch in length. His color was a beautiful, solid, shimmering red, and at maturity his tail had an arching elegant sweep that in profile made him appear like a carving in an Abyssinian bas-relief. He carried himself in a most unchickenlike posture; erect—a stone dropped from the end of his beak would have landed very close to the end of his toes. Even more striking than his color and carriage was his general disposition. He was the one animal I have ever seen that didn't know the meaning of the word *fear*. An eagle is fearless, but will still fly away from what is unfamiliar to him, and has a practical sense of what he may and what he may not do. This game cock, however, seemed to live in a world where fear, pain, and even pride had no meaning or did not exist at all. The sole emotion he seemed capable of can only be described as a specialized kind of contempt.

My rooster could not in the proper sense of the word have been called tame, but could be picked up any time. He was not tame; he simply didn't give a damn about anything. He would walk up to strangers and attack their legs with his spurs. He would attack dogs, cats, and, for a morning workout, would regularly thrash our terrified Rhode Island Red rooster. He felt he had a personal contract on anything alive that weighed less than two hundred pounds. In ways he was the most primitive raw-animal force I have ever confronted. And if his lineage could have been known, he perhaps represented three thousand years or more of selective breeding.

Game chickens are beautiful and admirable, but, frankly, they're a little hot blooded for either practical or ornamental purposes. The end of my rooster's days came when on about his thousandth attack on my father's legs, he found my father in one of his rare intolerant moods. As Mark Twain observed, he got hoisted by the neck. Very rapidly.

An attractive breed quite close to the game cock, however, the many different species of the homey chicken called *bantams* can serve a distinct service in a hobbyist's yard, even the yard of one who is primarily interested in a more practical breed or breeds. In size they vary from minute up the scale to medium-sized and display a full range of color and feather configuration. All tend generally to share the same common patterns of behavior. They do best when allowed their own free range, although it is much more convenient if they can be made accustomed to sleeping in the chicken house. This makes it easier to catch them, to encourage them to lay in one spot, and more convenient to pen them up when their free ranging becomes a nuisance.

While they do well ranging free, with a little attention they often become gentle and companionable pets. If given the run of the yard, they are the nonpareil of pest killers, a pest eradicator that leaves a residue no more harmful nor persistent than what can readily be scraped off your boot with a stick. However, given a free rein they can be quite destructive to a garden. They have a passion for lettuce and will devour it at any stage of growth. They are great scratchers and love to investigate any soft soil they discover; particularly the soil made soft by recent tilling and planting. It should be clear why it is advantageous to be able to pen them up when it's inconvenient to have them abroad.

As "working" chickens, bantams are clearly impractical. They are a pleasure to eat (when young) but their size is against raising them for meat. Their eggs are likewise small, but the hens turn out to be surprisingly prolific and long-lived layers. The hens have an unfortunate preference for making their own choices as to where they will lay their eggs. If luck doesn't step in, it takes persistence to locate a nest. Fortunately, whenever a hen lays she cackles, and a little crude sonar work, practiced over several days, may allow you to establish zeniths and home in on the nest. A wise old bantam hen may give up cackling, or cackle only when she is far away from the nest. Looking for nests is an absorbing pastime, but only small boys really have the time necessary to devote to it. Once you find a nest, always leave a couple of eggs, or the hen will abandon it.

Usually before a summer has run its course, every bantam hen will outfox you and make a nest you can't find. They are extremely

The Chicken at Large

good brooders and, when they do successfully hide their nests, will usually bring off a good-sized clutch of chicks. Of all domestic chickens, bantam hens are probably the best mothers. In many of the commercial breeds, in fact, the maternal instincts have been virtually bred out of existence. Although they may develop the desire to brood, it is nearly impossible to bring off a hatching of eggs with a White Leghorn, for example. She's too nervous and fidgety and just doesn't have the patience. Such older breeds as the Rhode Island Red, etc., are reasonably good brooders, but scarcely in the same league as bantams. In the past, many people kept bantams just for their good brooding qualities. They were regularly set with the eggs of larger chickens, and half a dozen bantam hens could do the work of a modest-sized brooder operation. Aside from their attractiveness and value as pest eradicators, it's worth keeping a few around just for brooding purposes. Certainly no hobbyist can resist eventually bringing off a hatch of chicks, no matter how his operation is geared.

Bantam hens may be absolutely neurotic in their dedication to setting. A neighbor had a bantam pullet of a most minute black strain which had improvidently laid a clutch of eggs on a narrow plank in the loft of his barn. One by one they rolled away and broke until finally she was covering only bare wood. Nothing would divert her fixed resolve, however, and finally—sadly—she simply died. There are always stories of near-legendary bantam hens credited with absolutely Homeric feats of brooding, such as a hen belonging to my grandmother which would go broody at the sight of a nestful of eggs. Virtually any time during the year she could be persuaded to take over a hatching of eggs, of any species and size. One of the values of this brooding fixation that is characteristic of the breed is that they can be induced to hover eggs of any variety and will remain on the nest much longer than the usual twenty-one days required to hatch a chicken egg. When I was a boy, we regularly used them to raise our meat chickens and replacement layers. We also regularly set one with turkey eggs for holiday birds, occasionally with duck eggs, and even with pheasant eggs.

Backyard Poultry Raising

There's nothing more fetching than a hen fussing about the yard with a clutch of baby bantams. If sometimes raising chickens, even on a small scale, begins to seem too mechanical and remote from the natural, allowing it all to take place on its own is psychologically refreshing.

Also, when you feel frivolous and impractical, there are all kinds of exotic eggs that are temptingly advertised in the various poultry catalogues which one can incubate under a co-operative bantam hen. Think of wild turkeys! Guineas! Think of geese! Swans? Definitely bantams must be given some thought. And they will help keep you from taking chickens too seriously.

Even a person who has not had the remotest connection with the farm certainly has some conception, even mental images, about setting hens, and even *setting a hen*, derived from nursery rhymes and fairy tales. Very few people have any notion of what you do to bring either about. It is possible to take a mature hen at random and talk her into being in the mood to set, but don't count on it. I have known farm women who accomplished this regularly, but I think it is an art that is hard to teach or be taught. The likelihood of success for most of us is so slender it's hardly worth the effort. But in the course of events it is normal for every hen, except those in which the instinct has been completely bred out, if she stays around long enough, to decide she wants to set, or "go broody." When you watch a broody hen, you quickly realize why the expression *broody* has been metaphorically attached to people who are in a certain crotchety frame of mind. A broody hen gives the impression of being both preoccupied and generally in a disagreeable mood. She clucks. Even though in the popular mind a hen always *clucks*, the only time one actually makes that sound is when she's broody. Broodiness usually comes on in stages, over a period of several days. At first she may appear in all other ways normal except that as she walks around she clucks. Gradually her temper and temperament deteriorate, and she absolutely glowers as she walks, with feathers ruffled up and showing a tendency to drop her wings and attack other chickens, or even her best friend, the farmer. From this inclination comes the country expression of a person or animal being

The Chicken at Large

"on the peck." When she reaches this stage, she's ready to set, or make the attempt. She may commandeer a nest whether it has eggs in it or not, simply flopping down in it and going to setting. If a bantam on the loose begins to be broody, you can be reasonably sure that she has a clutch of eggs stashed away someplace, and shortly you will be seeing little of her for the next three weeks.

If you want to try setting a broody hen on a nest of eggs, it's best to arrange separate facilities to see her through her term, something away from the hustle and bustle of the hen house. Incubating eggs is a private business, and it requires concentration. On the farm we used to set our hens in boxes filled with hay secluded in a shed, with food and water provided nearby. Occasionally there would be several hens setting there at once, and they didn't seem to disturb each other even if one hatched a couple of weeks ahead of the others. Unless you have a secluded or secludable building, it is handy to build a small hutch or cage for brooding. The facility need be only some three feet high by three feet wide by four feet long, containing a nest filled with straw or hay and supplied with feed and water. An A-frame-type structure is also suitable, and perhaps easier to build.

The first consideration in setting a hen is selecting the eggs. If the eggs from your own flock are fertile, you will probably use them, but you may wish to obtain eggs from another flock you admire, order them by mail, or obtain them from a commercial hatchery. Unless you care to go into controlled breeding, it is best to refresh the bloodlines of your flock occasionally with a fresh stock of eggs, or obtain a new rooster of proven lineage. When you select eggs, choose only large, perfect ones for incubation. Small eggs are apt to be infertile, or to hatch a sickly and unthrifty chick. Too large an egg may indicate a double yolk. These eggs, if they hatch, will yield a wildly abnormal chick. As a boy I periodically smuggled a double-yolked egg under a setting hen, and the chick(s) were sometimes multilimbed, multibodied, et cetera. They would usually die within a short time of hatching, or not be able to wholly make it out of the shell at all.

Begin by giving the hen a couple of days to get used to the hutch.

To get the hen started, first put a couple of eggs in the nest box. If she accepts them and settles down to incubate, add the rest late the following day. She will undoubtedly peck you furiously when you slip the eggs under her. I have never known the attacks to be fatal. A Rhode Island Red hen should be able to accommodate about eighteen eggs. When setting a bantam, it is necessary simply to gauge her size and guess how many she can hover. Give her a couple more than you think she can possibly handle. While the hen is setting, it is necessary to provide her with a constant supply of food and water. She will only get off the nest for brief periods once or twice a day, and, although she will probably eat little, she will drink a lot of water. When the chicks hatch, provide feeders and waterers for them as well as for the hen. Although the hen will be unable to

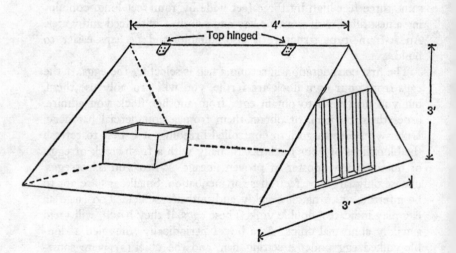

Simple A-frame coop for setting hen and her chicks, constructed of plywood, with "door" made of 1" × 1½" slats. No floor is needed.

The Chicken at Large

leave the pen, the chicks will not wander far from their home. After about two weeks the hen may be released if she and the chicks are provided a pen or space that is protected from adult birds. Except that the hen will take care of their warmth and devote herself to keeping them in line, the feeding and watering will proceed exactly as with baby chicks that you are brooding without the help of a hen.

It is also possible to buy (or incubate) baby chicks and give them over to a broody hen for adoption. The coop or hutch described above is also suitable for this. Situate the nest where you intend the chicks to be, and settle the hen in on a few eggs as if you intended her to bring off a hatch. She must be given a little psychological encouragement; although a chicken can't count days, they seem to have enough reasonableness to know that it takes more than a single day to hatch a setting of eggs. Therefore, let her brood two or three days. Then, at night, gently and surreptitiously introduce into the nest the chicks which you have purchased. The next morning most likely she will have accepted the brood and be extremely pleased with herself into the bargain. Since a hen can hover and look after more chicks than she can incubate eggs, you may introduce twenty or twenty-five chicks. It is also possible, where chicks are readily available, to give a hen that has brought off a hatching enough extras to bring her up to about twenty-five. Although you may have set her on eighteen eggs, it is foolish to count in advance the number of chicks that will hatch. You could almost make a proverb of that.

However, the pastoral image of having a hen wandering loose around the yard with a brood of chicks in tow is a hard ideal to realize. First of all, a hen with a brood does a great deal of scratching around, almost always in places you would least like scratched. Second, if there are fields of tall grass and weeds, it is very easy for the chicks to get lost, or to fall into ditches or holes that they cannot get out of. A bantam hen can come closest to managing a brood in tall cover, especially if the chicks are themselves bantams. The larger hens are not quite as good at it. And third, chicks are very easy prey to predators, especially cats. Cats can learn not to catch baby chicks, but it's their instinct to do it anyway.

While in most poultry operations, even with small hobbyists, the setting hen has gone the way of the horse-drawn plow, a brood hen still has certain definite advantages if you discount the time entailed. But as the old saying goes, "What's time to a hen?" If you have hens, access to fertile eggs, and are yourself possessed of the proper temperament, three hens can brood perhaps forty or fifty chicks and replace a considerable amount of equipment and attention. The reason I do it is because I think it's fun.

It is rather easy to become attached to certain chickens, especially if you raise only a small stock. If for no other reason than this, if you are raising chickens for meat it is wise to raise enough at a time that you don't begin looking at them as individuals. In this respect my wife has a steely resolve; she refuses to eat any animal with a name. If you are raising chickens for meat, you must avoid allowing this intimacy to occur. Even your laying hens will get to know you, will rush to you enthusiastically when you appear, and recognize the sounds you make before they can see you. Contented hens will sing as they stroll around in the yard, a very pleasant and homey sound, and this also endears them to you. For a small-time chicken raiser, therefore, it is both a pleasure and a psychological convenience to do a little chicken fancying on the side, in addition to your more practical and economical activities. A trip to the poultry barn at a county fair will introduce you to varieties, curiosities, and novelties of the poultry world that a novice can scarcely conceive. These fancy chickens are reared exclusively for pleasure (although they do lay eggs and are themselves no doubt edible). Again, I would recommend for starters the ornamental strains of the doughty bantam. You may lavish on your fancy chickens the affection it is impossible to give your laying hens or your flock of fryers.

The first chicken I ever called my own was a small, common bantam hen, when I was about four years old. She was the only chicken we had that did not stay in a pen and was as companionable to me as a dog. It became her habit to peck at the kitchen door to be admit-

The Chicken at Large

ted into the house, and go directly to my bed. There she would find a comfortable declivity and lay her semidaily egg. I always ate it scrambled. A complete and unstrained friendship with a bird is a special kind of experience, and its pleasures are impossible to describe, but wholesome and refreshing.

While for the most part the hobbyist will probably select on practical grounds one or possibly two breeds and pretty much stick with them, part of the fun of raising poultry is experimenting around and trying new breeds. In the beginning, it might be well to consult the county extension agent for advice on what breeds do well in your region. For starters, you will probably not go far wrong selecting from any of the standard dual-purpose egg and meat breeds if you wish to raise both meat and eggs. An all time favorite is the Rhode Island Red; only slightly behind is the Plymouth Rock. Both are reasonably large birds, the hens averaging four or five pounds. If these breeds are raised for fryers, at ten to twelve weeks they will weigh 2½ to 3 pounds. Such crosses as the Hubbard will gain somewhat faster and make appreciably more efficient conversion of feed. If you intend to raise ten thousand, these differences are important; if you raise fifty, the factors are of probably negligible significance. Other popular dual-purpose breeds include the White Plymouth Rock, or White Rock, and the New Hampshire. Their characteristics, aside from color and feather styles, are very similar to the breeds mentioned above. There are also several color and size variations of the Orpington breed, the most commonly known being the Buff Orpington. These are very comely birds, and enjoy areas of regional popularity. I have always had a soft spot for them because of their exceptionally attractive appearance. Since there is little real difference in performance in choosing among the standard dual-purpose breeds (and first-generation crosses between them), it adds dimensions of interest in chicken culture to change the breed of the flock from time to time, whatever your reason for growing chickens.

You will also find, particularly in the more remote rural areas, strains that seem to be more or less indigenous to that region or area. It used to be generally true, and is still true to a certain extent, that many marginally operated flocks of chickens were essentially mongrels. Such strains tend to be unthrifty either as meat producers or egg producers, and seem to be kept only because they have always been around. In *The House of Seven Gables*, Nathaniel Hawthorne describes such a strain of inbred birds, which he calls "the Pynchon Chickens." Although locally renowned for the exquisite flavor of their eggs, they laid few, and were in danger of becoming extinct because few of the eggs seemed fertile and the hatchlings seemed unusually weak and fragile.

But if breeds can degenerate through careless management, or no management at all, they can also be improved through careful management and selection of breeding stock. If you care to take the trouble, it is possible through selection to produce a private strain of a particular breed that is slightly superior in the qualities you choose to select than those commercially available. To embark on a breeding program, it is necessary to band each hen with an identifying tag and to keep a careful record of her life history, particularly a detailed record of her performance as a layer. A "trap nest" must be used to prevent error in determining her laying performance; this is a commercially designed nest which traps the hen when she enters to lay and holds her until she is released by the poultry man. The nest must be checked several times a day to avoid overlong confinement. When the hen is released, her identification is checked, and on her card is recorded whatever data you are interested in: when she began laying first, the fact that she laid this day, the size, shape, color, condition, etc., of the egg, matters of diet—whatever elements concern you and the ideal you are intending to breed toward. The hens of your flock will be evaluated in terms of such factors as beginning to lay early, large production, good quality eggs, and the length of the laying period, and the superior hens will be mated and their eggs incubated. It is another subject altogether, but breeders often use artificial insemination in controlled breeding programs. The process of breed improvement, or even breed sustaining, is obviously a time-comsuming one, but an engrossing and rewarding one for those interested in getting the most pleasure (hopefully, short of getting hung up on the hobby) out of raising poultry.

Actually the breeding standards of most suppliers of commercial chicks are so high that it is unlikely an amateur will produce a bird that is significantly, or even detectably, superior. But maintaining and improving the strain in whatever way has always been an absorbing concern of any animal breeder. If you have any intention of breeding your own even in a casual way, some systematic program for sustaining the breed must be part of the scheme of things or eventually everything will go downhill. If you would like to breed your own, but do not care to select your own breeding stock with the care and attention necessary to be a true "breeder," it is imperative that your breeding stock be sustained and constantly refurbished through the purchase of superior outside stock.

Whether or not he ever sets a hen, some time or other during a hobbyist's involvement with poultry raising he is going to have the urge to incubate his own eggs. A so-called "table-top incubator" (as distinct from the larger cabinet types) is not expensive and is easy to operate. They come with various capacities, and, although some are designed for a specific kind of egg, most will accommodate eggs from such diverse sizes as quail eggs to goose eggs. Home incubation will probably not result in appreciable economy, at least initially, unless several hobbyists with similar interests pool their resources in the purchase. However, if you discover that you wish to begin dabbling in more exotic strains of chickens (as well, perhaps, as other species of poultry), ultimately a home incubator may be justified as a necessity. Which is to say, you really *want* it. Many of the exotic strains may be obtained only by ordering fertile eggs from specialty breeders. Even if such a breeder is sufficiently close at hand for mail ordering, home incubation certainly helps resolve questions of live delivery.

Incubators may be ordered through the various poultry catalogues, through feed stores and poultry dealers, or through the farm catalogues of the large mail-order houses. Not infrequently a used one may be picked up at a considerable saving in price. They are not difficult to operate, and each will come with complete operating instructions. It is necessary to offer the reminder, however, that occasional attention is required throughout the incubation process. It is not simply a contraption you can turn on and return twenty-one days later to find it filled with fluffy yellow chicks. They are not complicated, but they are not totally automatic either. As a boy I once tried to design and build my own incubator. The heat source was a kind of lamp I had made from a shoe-polish bottle and strand of sash cord, fueled with radiator alcohol. It baked the eggs and very nearly burned down our house.

Other Poultry, Excluding Rabbits
<<<<<<<<<<<<<<<<<<<<<<<<<<<<<<<<<<<<<<

Of all poultry the backyard hobbyist may try, chickens are certainly the most practical, versatile, and probably the most rewarding. But if you like livestock, and birds in particular, if you have a restless urge to putter around and see new things happen, and if you are fortunate enough to have the room, sometime you will find yourself irresistibly drawn to raising something besides chickens. Of the nongallinaceous fowl, I have found the easiest, and in many ways most pleasurable, to be ducks.

The first ducks I ever raised were mallards. As a boy, I had pilfered the eggs from a wild duck's nest (the statutes of limitation have long since run out, permitting this confession, and I'm honestly sorry and would never do it again) and set them under a bantam hen. The eggs were large, and she was a rather diminutive bird, and to hover them she had to spread herself until she was shaped like the rug in front of a small fireplace. Within a few days after their hatching, the ducklings were so large that when she hovered them she was hardly able to touch her toes to the ground. All of those little black beaks flanked by shining little black eyes protruding through her feathers gave her the appearance of being the product of a runamuck Russian biologist. She found her ducklings a terrible trial and when they went for a swim in our irrigation ditches she would sit on the bank and grieve. Whether it was because it never occurred to them to fly, or because they grew too fat to fly, they never achieved any flight other than a brief sort of fluttering above the surface of our ditches. I have always attributed their failure to overeating.

A neighbor had a small flock of domestic Pekin ducks which lived in a moderately large pen, and a wild mallard drake began courting a nubile young hen. When he dropped in to court, he also made use of the opportunity, as courting males have been doing from time out of mind, to raid the family larder. Eventually his stays became longer until, finally, by the time she had hatched a clutch of remarkably variegated ducklings, he was too well fed to be able to fly out of the pen at all. I have thought of that old drake often, as a kind of almost biblical parable. There's a political, social, and personal message there we can all respond to.

Ducks are very easily raised on a sort of let-alone basis. You have to have room, though, to allow them to roam at large and forage, and a little tolerance as well. Ducks can be very messy. They also are voracious eaters of vegetation. This means that with a little room to forage, they are virtually self-sustaining through a major part of the year. It also means they will raise havoc with a garden if you raise one. They prefer their vegetation wet, and you may find after a rain that your growing corn, which previously they had seemed not to notice, has been grazed down to bare stalks. Any-

Other Poultry, Excluding Rabbits

where you leave water running, they will make a puddle, usually to your disadvantage, if not to your actual distress.

The duck seems to be one of the more intelligent of the barnyard fowls. Particularly if you have only one or two, you will discover they have a remarkable kind of individuality and personality. A hen duck (that's really a redundancy: a duck *is* a hen; a *drake* is a male) we had for several years became one of the most appealing animals on our place. She had an extravagant fondness for my wife and would flop down in front of her and spread her wings, begging to be picked up and caressed. Jessie (that was the name our duck went by) loved table scraps and would position herself on our lawn and fix our dining room window with an unwavering eye when we were eating, waiting for clean-up time. When she had selected and eaten her choices from the table scraps, she allowed the dog and cats to have a turn.

In her first spring as an adult she established a nest in a doghouse and began setting. I felt sorry for her and the futility of it all, wasting her energies incubating infertile eggs, so one day when she was off the nest I disposed of the eggs. Shortly she began laying again, and when she had enough eggs to satisfy her, again went to setting. This time we managed to rustle up a dozen eggs from a farmer with a barnyard full of assorted poultry, and all hatched. Two did not free themselves from their shells until late the second day of hatching and seemed a little weak, so I brought them into the house to gain strength. The next morning they began quacking frenetically, and apparently Jessie heard them. She began marching around the house with the remainder of her brood following, for all the world like a feathered but indignant Joshua and his troops, waiting for the walls to tumble. Needless to say, I gave her back her ducklings, and she stopped marching.

There is absolutely nothing about a baby duck that is not appealing. The pleasure of watching a duck with her brood (at least one time around) is probably more than equal to their harm to the garden. Very skillfully, she will spread them like a platoon of infantry deployed as skirmishers and sweep across a lawn, raising clouds of small bugs before them. For a time we had a bad infestation of lawn moths, but after the year of the duck, we were never troubled again.

We kept a drake from Jessie's hatching, and for several years she would regularly bring off two broods of ducklings every summer. If one kept all that hatched for breeders, the geometric progression of successive offsprings and their offsprings would make calculating machines tremble. In short order the earth could be covered with ducks up to the world's highest belly button.

Jessie's hatches would regularly number a dozen or fifteen. As she grew older, she tended to make her nests some distance away, often as much as a quarter of a mile. Every afternoon, however, she would come home and expect to have a puddle made for her. During her life she became very adept at making her specific wants known, imperiously. She would also expect to be fed—the only time during all but the severest winter months she did not disdain anything except table scraps she had not foraged for herself. Apparently she picked the afternoon to leave her nest because it was warm enough that the eggs were in no danger of chill being left without her body heat. She might be off her nest only fifteen minutes, or if it was quite hot, as much as an hour. But the last thing she would do during her break was to go into the puddle and deliberately and carefully fluff her feathers and saturate them with water. From both observation and consultation with others, I discovered this to be an instinctual and necessary precaution. The moisture in the feathers is transmitted to the eggs, and helps keep the shells viable through the incubation period. It is the duck's natural way of taking care of the humidity problem, a major technological concern of those using electric incubators.

During the dead winter months, Jessie demanded to be fed. Her favorite feed after table scraps was dried cat food. She would eat grain or poultry food if it was given to her, but she always let you know she was displeased by it. Reliably, you could go out any time of the day and call "Kitty!" and get a responsive quack from somewhere on the place, and shortly herself would heave into sight, waddling full bore.

Other Poultry, Excluding Rabbits

There was a distinct officiousness about her general bearing. I've noticed that quality in ducks generally, but she was a master of making you feel somewhat *de classé* and distinctly of less importance than herself; that somehow her business was of considerably greater significance than your own. It may have been imaginary, but she put across the impression that she was the organizer and manager of the activities of all the outdoor beasts. There was a general languor about the animals until Jessie appeared among them, then suddenly everyone seemed to know what to do, and deliberate and meaningful activity replaced bored inaction. Even our horse, a lovely quarter horse mare whose sole interest normally was to throw me from her back and hurt me, seemed to develop other and more benign concerns when Jessie took command.

But to be practical rather than rhapsodic, ducks can be raised with much the same equipment and in essentially the same facilities as chickens. It is not even absolutely necessary that they be provided with water for swimming, although they decidedly enjoy their pool. Ducks mature rapidly, and are generally considered as prime roasting ducklings at about the age of eight weeks, when they will weigh about seven pounds. That is faster and heavier growth than frying chickens. If you have facilities for them to forage, it is possible for a useful number of them to reach this weight on a very small actual outlay for feed, although they will mature faster if given supplemental feed.

If you begin with newly hatched ducklings, whether you incubate them yourself or buy them commercially hatched, as soon as you move them to their brooder facility, they should be provided with food and water. They will need heat for the first four weeks, or the first two or three weeks in warm weather. Begin them at a temperature of 85–90 degrees the first week, lowering the temperature by about five degrees each succeeding week. Like baby chicks, ducklings need a clean litter on their floors. Change the litter regularly to prevent accumulation of dampness. Although aquatic birds, initially they are very sensitive to moistness and somewhat subject to pneumonia. When the ducklings are about four weeks old they should be well feathered out, and if at all possible, by this time, they should be provided access to an outside run. A pond is not necessary, but they will certainly enjoy a swim. Lacking other facilities, I have used a small child's plastic swimming pool. However, it will need to be cleaned almost daily.

Commercially prepared feed mixtures for ducks are available, and it is preferable to feed ducklings food in pellet form rather than mash. Starter pellets should be about ⅛ inch in diameter, and should be fed for the first two weeks. Change then to a grower mash, about 3/16 of an inch in diameter, until they have reached the size you would like for butchering. If they are given access to forage, these pellets can be fed in supplement to the food they obtain on their own. Keep watering facilities close to the feeders because as ducks eat they have immediate need for large quantities of water. And of course if the ducklings are running loose, either at liberty or in a pen, make certain that they have access to water outside at all times.

Ducks are dressed much like chickens. They may be decapitated, or they may be killed by the knife method. They do not dry-pick as readily as do chickens killed by the knife method, however. Plucking either ducks or geese is an aggravating prospect. One method is to scald the bird in hot water (140–150 degrees, slightly warmer than for scalding chickens), containing detergent or shampoo. The detergent helps dissolve the oil in the feathers, allowing more efficient penetration of the water. After dipping them until the wing feathers loosen in hot water, they should be rinsed thoroughly in clear warm water to remove the detergent. Otherwise the feathers will be slippery and difficult to get a grip on.

Probably the best method of all for removing the feathers from either ducks or geese is to dip them in melted wax. A goodly quantity of wax is required, and it must be melted in a pot or container large enough to submerge the bird. Immediately after, douse the bird in cold water until the wax sets up, then peel wax, feathers, and all from the body. The wax can be reused, melting it, and straining off the feathers. If the wax system is used, by all means remember that the wax is highly combustible and take appropriate steps for safety. The wax method should leave the body quite clean. But if it is scalded and plucked, there very likely will be pinfeathers remaining. As in the case of pinning chickens, a dull knife is a useful tool in helping to remove them. As a final step, the body may be singed to remove the small hairs which remain.

Once the bird has been plucked, individual preference determines the next step. Some (myself included) prefer to draw and dress the duck in exactly the same fashion as dressing a roasting chicken. Some, however, prefer to age the bird with the viscera still inside. One method is to store the bird twelve to twenty-four hours in ice slush, undrawn; that is, with the viscera unremoved. Birds (chickens too) may even be marketed this way. This is the so-called "New York dress." One also hears of "hanging" the birds, a system preferred in some areas and some countries for all birds, but especially waterfowl and game birds. They are hung by the legs, with adequate ventilation, usually outdoors, uneviscerated and usually with the feathers still on. Depending on taste, custom, or weather, they may be hung for a day or two to a week or more. This is not something I recommend, or pretend to have specific information to pass on about, but it is a practice most have at least heard about. One also hears of an oriental preparation called "wind duck," in which the duck is hung high until it is actually dried before being eaten.

While chickens are a dual-product bird—meat and eggs—North Americans have never cultivated a taste for duck eggs. Their flavor is a little more pronounced than chicken eggs—hard to describe, but definitely other than "bland"—and the white has a firmer consistency than the white of a chicken egg. Some nations and ethnic groups eat them avidly, in decided preference to chicken eggs. Sometimes duck eggs will be found in markets about Easter time; they are superlative for Easter eggs. They have a smooth greenish-white shell that is quite hard, they take dye very well, and they are large enough to allow an egg dyer room for expansive, creative work.

During the spring laying season ducks will produce eggs quite prolifically. Even a very few ducks will lay more eggs than any but the most enthusiastic hobbyist will wish to incubate, certainly more than the hens could set on. It is wise to think about possibilities for consuming them. If your palate does not develop a taste for duck eggs, they may nevertheless be used without noticing the difference in all kinds of cookery. Also they may be frozen, the same as chicken eggs, and used later for cooking purposes.

As with chickens, there are a number of species of ducks to be considered when choosing what to raise. In the United States, the White Pekin would have to be called the favorite. Not unexpectedly, the breed originated in China. It is a large duck, males running about nine pounds at adulthood, females about eight pounds. They are also good egg producers, which is important for one who wishes to breed his own ducks and hatch the eggs himself. I have read that the Pekin is a poor brooder, but my experience with the breed has been otherwise.

The Aylesbury duck is much like the Pekin in size and productivity, but is much more popular in England than in the United States. Many farm flocks consist simply of tame mallards, the wild strain which has been raised in captivity for many years. Most find mallards fine nesters, but their smaller size makes them generally less desirable than the standard domestic strains. For a hobbyist, their chief attraction is that they are pretty and look authentically wild. The Rouen duck closely resembles the mallard in general appearance, but is appreciably larger. Their voice is also softer than other breeds of ducks. The Muscovy duck, which despite the suggestion of its name originated in South America rather than Russia, is considered a very good meat producer. It is distinguished by a large, red, rather fleshy unfeathered area around the eyes, although it otherwise resembles the Pekin. Among the domestic strains, they receive high credits for nesting, but are not especially good egg producers.

In addition to the meat producers, there are breeds of ducks which have been developed for egg production. Probably the best of these is the Khaki Campbell. They are rather small, which limits their value as a meat animal. However, they are extremely prolific

Other Poultry, Excluding Rabbits

layers. In tests they have been clocked at a production rate of 365 eggs per laying year, which is a figure the best laying hens have never approached. If ducks eggs ever become commercially popular in this country, the Khaki Campbells will be the White Leghorns of the duck world.

By and large, hobbyists who become interested in raising ducks will be content with a small-scale operation, more attracted by the pleasure and curiosity of it than practical advantages. A drake and a duck or two (although ducks prefer to be monogamous) should comprise an adequate-sized flock of this specialty poultry. Unless you have, or wish to cultivate, a taste for duck eggs, or plan on making a real bash at Easter time, most certainly a meat-producing breed would be best. In spite of cautions I have heard about their nesting capabilities, my own experience with Pekins makes them appear in all respects a good choice, even for self-nesting.

Given the availability of adequate space, everyone will one time or another feel the necessity of raising geese. Their apparent size—they are large, but not actually as heavy as they look—makes them an eye-catching addition to the place that has room for them. If there is a pond available, I would consider them as almost indispensable. The sight of their placid cruising on the surface of a still pool of water is as tranquilizing on a summer afternoon as a cool can of beer. With room to forage, geese may be totally self-sufficient after their first four weeks or so. They love vegetation of almost all kinds, including plain grass and plain and fancy weeds. In the fall, confinement with a feeding of grain to fatten them up and mellow their flavor will put them in prime condition for eating.

The fact that they will not eat *all* vegetation has given them a peculiar ecological niche in certain types of agriculture. A large number kept fenced in with certain crops they do not eat will serve to keep that crop virtually weed free. I have heard of them used to weed cotton and strawberries (before blossomtime) and have seen them used extensively in weeding peppermint, spearmint, and asparagus. Sometimes geese are actually rented to farmers for the summer or through the growing season, much as fruit orchardists rent hives of bees for spring pollination. Mint and spearmint impart a disagreeable flavor to the flesh of geese which have spent the summer weeding, and, if they are to be eaten, they need several weeks on corn or other grains to fatten and "cure" before they are fit for the table.

Although they require a great deal of water for drinking purposes, it is not absolutely essential that geese be provided with water for swimming. They love to swim, of course, and are never more graceful and lovely than when placidly afloat, but the swimming pond is not absolutely required to raise them. I have heard some say that geese will physically mate only if provided water of swimming depth, but in my experience they have mated perfectly well on dry land. Breeding geese, however, can provide peculiar problems that do not occur with other livestock. Geese have their own decided preferences for the lines their social and domestic lives will be conducted on. In fact, in these respects geese are so similar to humans that an institute in Bavaria has for many years sustained a program for studying the domestic lives of the wild gray goose precisely to improve the understanding of human family and social relationships. Geese are essentially a monogamous animal and in the wild often, or even usually, when one of the partners in a pair is killed, the other will remain solitary and celibate for the remainder of its life. Geese are very tender and solicitous to their offspring, and family groups develop elaborate interrelationships which are undetectable to the inexpert human eye.

Therefore the pairing of geese for breeding purposes is not always an obvious and easy thing. The casual transfer of mates can be as resented as the casual transfer of human matings, if not more so. If he has a mated pair, in most respects the hobbyist is best to respect that pairing, and either to incubate their fertile eggs or allow the goose to do it for him. Or else simply to buy eggs or goslings in the spring if he intends only to raise a few each year for meat or weeding.

The two most familiar breeds of geese are probably the gray Toulouse and the Embdens. The breed of the Toulouse probably originated several millenniums ago through the domestication and subsequent selective breeding of the wild gray, or gray-legged goose. They are large, dignified (if ungainly), and perhaps a little dowdy and Hausfrauish. An old and established goose of this breed begins to take on the same kind of apparent domestic permanence of an old dog. Indeed, they are probably much longer lived than most dogs. Geese are among the longest lived of the wild fowls.

The Embdens are about the same size as the Toulouse (though they may appear to be smaller) and are a very attractive white bird. Seen swimming at a distance, the uncritical eye might well mistake them for swans. Another domestic breed, the White Chinese, is slightly smaller. Unless you make the serious decision to go into geese in a big way, your first selection of breed will probably be determined by what is available.

If you begin with goslings, start them off on a commercially prepared mixture. There is a commercial feed mixture available for virtually any animal you might choose to raise. Besides special feeds for the various breeds of poultry, there is pig chow, horse chow, monkey chow, etc. When I discovered that there was also a commercial trout chow, I concluded if you took a whim to raise duck-billed platypuses, you would have only to go the the feed store and ask for platypus chow and your feeding problem would be at an end. But offer your goslings greens when they are a week or so old, being careful at the same time to keep them constantly provided with grit. Within a month, greenery will probably be their chief diet. If you have facilities for them to run, they will probably forage for themselves and require no further feeding until autumn, when you will want to finish them on grain before butchering.

The very young goslings are quite vulnerable, and for their first couple of weeks must be provided warmth and be cared for in exactly the same fashion as ducklings. Like ducklings, they must also be protected from dampness for the first couple of weeks. They soon become extremely tough and when they begin to forage are one of the most carefree varieties of poultry you can raise.

It is rather otherwise with turkeys. Young turkeys are extremely delicate, and can require an exasperating amount of attention. Fortunately, the problems which arise when raising turkeys are considerably lessened when you raise only a few. Domestication and selective breeding have made them a far different creature from their wild ancestors. The wild turkey is rated among the brighter of the birds, wiley, tough, most self-sufficient. The modern livestock turkey has been bred for efficiency in feed conversion and a carvable breast, and the virtues of their wild ancestors have been lost. The rather unnatural anatomy developed in the commercial turkey makes even normal breeding difficult. Mechanical assists for mating are resorted to, such as little canvas saddles for the hens to wear. I can remember when farm turkeys routinely incubated their own eggs, but I suspect that by now it would be almost impossible for most present-day strains to do so.

Fortunately, the hobbyist who raises only a few—and who really needs more than a few?—is at a distinct advantage over those who raise turkeys commercially. Some of the problems turkeys present to their growers results from the large numbers being raised together. The small operation is much less likely to be troubled with disease, cannibalism, smothering, or stampeding, to name only a few of the harassments turkeys give to commercial growers.

One of the first rules of raising turkeys is to keep them separate from chickens, or where chickens have been raised. This rule I have broken and have seen broken over long periods of time, when only a few birds were being raised. But take this caution: If you raise them around chickens or use facilities previously used by chickens, you run the risk of disease.

On the farm where I grew up, we regularly incubated turkey eggs under a bantam hen. When the eggs hatched, the bantam hen had the patience to tolerate and attend to the helplessness of the young turkeys, or *poults*, as they are properly called. In short order they would become as big as, or bigger than, their foster mother and would hoist her clear off the ground trying to get under her. This was certainly the most trouble-free way of raising turkeys, leaving the bantam hen with all the worry and letting her do all the

Other Poultry, Excluding Rabbits

work until the poults were large enough to pretty much make it on their own.

Realizing the chanciness of dealing with turkey poults, and accepting the possibilities of loss or failure, a stalwart amateur may well wish to try his hand at raising half a dozen or so just to show that he's made of good stuff. If you buy day-old poults, they will have about the same brooding requirements as baby chicks. Feed them a commercially prepared turkey starter. In the beginning you may find that you will have to teach them to eat and drink. Put their mash in a shallow box, and strew a number of glass marbles in with the mash. The shininess of the marbles attracts the poults to peck, and when doing so they will accidentally get an occasional beakful of food. Eventually they will catch on and eat intentionally. To help them learn to drink, dip the beak of each into their water, carefully, to avoid dampening the poults. It may, or will probably, be necessary to do this on several occasions before they all learn where the water is and what to do with it. If brooding day-old poults, be scrupulous about cleanliness; use an absorbent litter and change it frequently.

Feed the poults starter mash for the first eight weeks. The starter mash fed to turkeys is considerably higher in protein content than the chick starter mash—up to about 28 per cent. At eight weeks, change to a pelleted growing mash, which should consist of 20–22 per cent protein. At this age they should also be introduced to grain, either whole or cracked. Provide the two feeds in separate facilities, and give them as much as they want of both. When whole grain is fed, they must also be provided with a constant supply of grit. If it is available, they should also be provided green feed—grass, lawn clippings, garden scraps, etc.—after eight weeks. The smaller breeds of turkey are usually ready for consumption at about twenty-two to twenty-four weeks. A June poult, in other words, will make a fine Thanksgiving turkey, or a stupendous Christmas bird. They may be butchered and frozen at any desired size, but a fresh turkey is a treat that should certainly not be missed after going to the trouble of raising your own.

If you have facilities for allowing your turkeys to range, and if the weather is suitable, they can be turned out at about four to six weeks of age. It is a definite health advantage if they have access to sunlight and green food, and the more forage they consume the smaller the investment in feed in the finished turkey. They must continue to be provided shelter throughout the entirety of their growing life, and you will very likely find it necessary to remind them to go inside at nightfall. Large flocks of turkeys are quite skittery, and totally given over to the basest impulses of the herd instinct. They will stampede at an unusual noise, and in a large operation they will pile themselves against fences, buildings, into corners, and kill each other in great numbers. A half dozen turkeys, however, find it difficult to bring off a particularly thrilling stampede regardless of the source of the terror, although they will race around with the mindlessness of grade-school children just released for recess.

The two main varieties of turkeys are the bronze and the white. There are a number of subvarieties of both, distinguished chiefly by mature size and body configuration. Some have been developed to have such huge breasts that they find it difficult to move around (although presumably they can still stampede in fine style).

Mature birds are butchered in about the same way as chickens. It is much more convenient, however, to hang them by the feet and kill them with a knife. The traditional hatchet has a picturesque appeal, but wrestling a twenty-five- or thirty-pound indignant and alarmed turkey on to a chopping block hardly leaves a confident hand free to chop or a clear eye free to aim. Turkeys may be dry-picked if killed with a knife, but probably the job will be easier if you scald. At best, picking a turkey is time-consuming. Every part of the dressing of the bird is complicated by the sheer size of the animal. However, when you eat a home-grown turkey, really fresh, you will quickly forget the annoyance or exasperation he put you through in raising him and the particular aggravation of his final day.

Another specialty bird, so common as to be overlooked, is the pigeon. They may be raised in the simplest of facilities—the loft of a

Other Poultry, Excluding Rabbits

barn, or garage, a dovecot in the attic, in the backyard, or whatever. Pigeons can be raised where facilities for other kinds of poultry are inadequate or completely out of the question. They are eaten as squabs, young birds from twenty-five to thirty days old. At this age they have completed feathering under the wings, but have not yet commenced to fly. A breeding pair of pigeons may produce ten or more squabs per year, two at a time. Once established in your facilities, pigeons become almost totally self-sufficient.

In the Middle Ages, great quantities of squab were consumed. In fact, the pigeon population was a constant source of contention between landowners and their tenants. Vast pigeon-breeding facilities were maintained at the manor house, and the birds ranged free across the estate. Because of their numbers, they frequently ravaged the crops of the farmers, but, by the terms of their leases, the crofters were forbidden to molest the birds. Essentially, that is what most who raise pigeons for meat do today, allow the adults to forage from the fields of their neighbors. Hopefully you will not risk bringing on another peasant revolt by abusing the privilege.

The common rock pigeon, the pest of cities, is one breed sometimes kept for squabs. However, larger and more attractive breeds are available and preferable and will make more efficient use of your neighbors' produce. Such breeds as the White King and the Giant Homer produce a squab considerably larger than the rock pigeon. The young of these breeds should weigh from fourteen to fifteen ounces at harvesting time, versus eight to ten ounces for the rock pigeon. Pigeons fanciers are themselves a numerous breed, and the periodicals and books they have spawned, devoted to the subject of pigeons, are legion. More information is available on pigeons than any but the most devoted would care to know. Even if you are not interested in raising squabs for food, a few of the ornamental varieties are a pleasant addition to a home. When you feel the need to look at a bird, a pigeon is quite suitable and always available. A strutting Fantail or Pouter, or the antics of a Tumbler or Roller, will soon dispel the prejudice the park pigeon has earned for the general species.

An interesting, lovable, and exasperating bird everyone should try, if the facilities are available, is the Guinea. The name immediately conjures up the fabled culinary specialty, "breast of Guinea under glass." They are a superlative meat bird, perhaps the best flavored of all domestic fowl. They are modest in size, considerably smaller than their live appearance would give you to believe. They come in various colors, but the commonest is a slate color, patterned with numerous white specks. They have an uncomely head, which seems disproportionately small for their body size, crowned with a sort of helmet of red, with bluish splotches. You learn to think they're lovely.

The only disagreeable quality of Guineas is their call. Actually they have two calls. One transliterates as a strident "Tough luck! Tough luck! Tough luck!" Some people find that depressing to their morale as well as irritating to their hearing. They make that sound principally when they are alarmed. The other sound is one they make when they are *thinking* about becoming alarmed. It defies transcription, but is a kind of strident metallic squeal, with excellent carrying quality. It sounds quite like a defective bedspring. They tend to raise a cry when anything unusual occurs, and they define the term *unusual* very loosely. It is said that they are superlative watchdogs. They will indeed give out an alarm if a stranger appears, but to a Guinea everyone is a stranger.

Guineas never consider themselves domestic, although they definitely become attached to a place and seem to feel easy with individual familiar people. They are marvelously self-sufficient foragers, and seem on the whole to prefer what they scrounge to what you offer them, in all but the leanest months of the year. It is difficult to keep them in confinement unless you have a pen that is completely enclosed. They can and do fly with ease and proficiency and are oftener seen on top of sheds than in them. When you walk into a flock of Guineas they will swirl up and around you like the agitated Ping-pong balls in a bingo machine.

Given any opportunity, Guineas prefer the life style of a wild bird. They easily become feral, and I have seen large flocks of them living completely wild. During the summer, Guineas will more often roost in trees than inside their appointed lodgings. When laying sea-

Other Poultry, Excluding Rabbits

son arrives, they like to hide their nests, and do it very skillfully. If you find the nest and wish to remove some eggs, you may safely take most of them, but not all. If you take them all, the hen will transfer her business elsewhere. If you have an incubator or a setting hen (bantam), that is probably the most efficient way to raise them. Guinea eggs have unusually tough shells. It is a common saying that you can throw one in the air as high as you're able, and it will not break when it hits the ground. I have never tried the experiment. Partly because of their habit of bringing off a hatch in deep weeds or heavy covers, Guinea hens on their own tend to bring only a small part of their hatch to maturity.

Baby Guineas—their proper name is *keets*—are commercially available through mail order and specialty poultry hatcheries. They are frightfully expensive. It is probably best to begin with adult birds if you can find them. Keep them penned up for several weeks after you get them, or they are apt to take off on their own. They have been known to return to their original home over quite a distance if they have not had a chance to become acclimatized before they are released.

Guineas are a most difficult bird to sex. I have been told that the males will always have a few white feathers in its wings. That sounds very unscientific, but there are always a few Guineas that have white feathers in their wings. It is also said that only the hens say "Tough luck!," that the males make only the squeaking-bedspring noise. Since the hens also make the second sound, this test is maddeningly difficult to administer and, for this reason, ultimately rather unreliable. Guineas are always either all sounding off or all silent, and with even a small number the confusion is demoralizing if you're trying to listen to what one particular Guinea is saying. The wattles of the males are slightly more pronounced or longer than the wattles of the hens. You have to have looked at a great number of Guineas, closely, before you become an accurate judge of their relative sizes. Frankly, I mistrust and suspect anyone who claims accuracy in determining the sex of Guineas.

My experience has been that if you wait around long enough, someone will eventually give you a pair of Guineas.

Chickens Cantankerous and Droopy

There's a hoary old gag about the farm lady who owned only a hen and a rooster. After a period of time, the rooster got sick, so she made chicken soup out of the hen to nurse him back to health. Perhaps it didn't hurt.

All varieties of poultry are subject to a wide range of different ailments, many of them fatal. The more crowded the conditions under which they are reared, the more vulnerable they become to various infections and maladies, and the more epidemic infections will be if they occur. Fortunately a great many of the poultry diseases have been effectively brought under control by breeding and medication. The hobbyists had best seek professional diagnosis and advice when disease is suspected.

If you obtain your chicks commercially, they should be certified as free from diseases which are transmitted from the hen to the egg through viruses. Beginning with certified chicks, for the first year or two a small chicken raiser is unlikely to experience any trouble with diseases at all. To keep things that way, prevention is far and away the best approach. Before you bring in your first chicks, have the facilities cleaned sufficiently in advance so that everything has had a chance to dry out and the fumes of disinfectant to dissipate. Once you have installed the chicks, change the litter frequently, and thoroughly clean the waterers every time you fill them. To keep the chicks from dampening themselves, which can lead to pneumonia, make certain that the areas around the waterers do not become soggy. As soon as you move the chicks from the brooder area, immediately clean and disinfect the whole area and all of the brooder-house equipment. This is particularly important in preventing infestation with internal parasites.

In a working poultry shed or chicken house, remove the droppings and litter frequently. If you have a droppings pit under the roosts (see "Housing the Humble Chicken"), it will probably not have to be cleaned out more than three or four times a year, depending on the size of your flock. Surroundings that are clean and tidy are a distinct deterrent to the development of wide varieties of health problems.

Proper feeding is vital in preventing disorders which are associated with nutritional deficiencies. Feed mixtures will tell on the labels on the bags both the nutritional content and any medicative additives. Chickens may also be vaccinated against several diseases. Like humans, chickens are healthier and more disease free when given adequate opportunity for fresh air and sunshine, and are made more resistant in general if their diet includes greens.

If you raise broilers and roasters, they will have run their life span before major infestations of most common diseases could develop or cause serious problems. Your chief guarantee that the chickens will

Chickens Cantankerous and Droopy

develop no problems that are essentially inherited, or come from virus transmission through the egg, is to buy certified chicks. You may want to vaccinate pullets being raised for layers against four potentially troublesome respiratory ailments: Newcastle disease, bronchitis, fowl pox, and laryngotracheitis. It is best to consult your county agent and consider his advice as to whether a complete vaccination program is desirable in your locality and with your scale of operation. Whether or not to vaccinate is a question to consider early when raising pullets, for the vaccination must proceed on a time schedule, beginning early enough to complete the cycle by the time they are around eighteen weeks old and before they begin laying. Frankly I have had no disease problems in recent years raising chickens, and suspect the small number of birds I keep, plus the access to sunshine and green feed available to them, contribute to this happy situation. I believe few hobbyists nowadays have serious problems with disease. If you do have trouble, or suspect you are having problems, consult your county agent or a vet.

Vermin may be troublesome, particularly in warmer climates. The northern states have fewer problems with worms than the southern states where they constitute a plague. In a frost-free area, flocks continue to be recontaminated by eggs which survive and repeat their cycles in the soil. The best approach is to prevent infection by bringing the chicks to clean surroundings and keeping everything clean, removing all droppings promptly and disposing of them in an enclosed pit or compost heap. Medication for worms is available, but it is much better to prevent them from becoming established. Once worms become established in a chicken yard in the warmer states, eradication is very difficult.

Other vermin which may be troublesome are mites and lice. Mites, usually called gray or red mites (they become red when filled with blood), are tiny arachnids which stay on the body. They will commonly be located under the wings and around the vent. Lice are insects which stay in cracks, etc., coming out at night to feed. Since both may be carried by sparrows and other wild birds, there is constantly the threat of possible infestation. Control of mites is best achieved by painting the roosts and walls of the chicken house with a commercial mite killer. Lice may be controlled by painting the roosts with nicotine sulfate. In the realm of folklore, I have heard that if the roosts are made from certain woods, especially of sassafras, there will never be trouble with lice. I have never tried this method, but if I lived where sassafras grew I would like to experiment with it.

An occasional case of paralysis may be encountered in a flock. This is a genetic ailment, and there is nothing you can do to cure it. It shows up more or less at any part of the growing period, but almost always before the birds are fully mature. The symptoms are difficulty in walking, vision problems often leading to blindness, and ultimately a complete inability to use the legs. Affected chickens should be disposed of as soon as the symptoms appear. Lameness associated with twisted feet will frequently show up in birds being raised for meat, but has nothing to do with paralysis. If the eyes appear to be normal; and the symptoms exclusively local and visibly physical—i.e., bent legs and twisted feet—the disorder is not paralysis. Except in rare extreme cases, birds with these malformations will mature at a more or less normal rate.

Baby chicks are subject to minor disorders that are fairly simple to remedy, but if ignored will lead to their death. Chilling, for example, may result in a lethargy or general droopiness. If the chick is isolated and kept warm, dry, and free from drafts, often it will recover in a few days. Sometimes the vent of a baby chick will become blocked by excrement sticking to the feathers. Soften the matter with a tissue soaked in warm water or mineral oil, and gently work it loose. There is probably a name for this disorder, and its treatment as well, but, if so, I have never heard of them.

One of the most aggravating maladies of chickens is one which they totally bring on themselves: cannibalism. Baby chicks and baby turkeys particularly are prone to this abhorrence. The presence of a sore or small injury, even a dirty spot or off-color feather, may attract the pecking of the other chicks. In short order, the mob can kill a chick, or leave it with large bleeding wounds. If you find a chick that is being pecked, remove him immediately and put him in a separate pen. Commercial preparations are available to spread on injuries, which both promote healing and inhibit further pecking. It is said that chicks raised under infrared heating lamps are less prone to cannibalism than when other sources of heat are used.

Probably the surest way to keep cannibalism from occurring is by use of a process known as "debeaking." It is much less barbaric and mutilating than the term suggests. When you order your baby chicks, you may specify that they be debeaked at only a fractional additional cost. Or you may easily debeak them yourself, using fingernail nippers. Simply trim the sharp, downward-hooking part of the upper beak as you would clip one of your own nails. Like a fingernail, the beak will grow back in time, and unless you take a deep cut, the difference between the trimmed, after it grows out, and the untrimmed will be virtually undetectable. More radical debeaking removes a larger portion of the upper beak with an electrically heated device. Debeaked chicks as well as debeaked adult hens are slightly less efficient feeders, being unable to pick up spilled scraps of grain or mash around the feeder. While they will waste some feed, the awkwardness in pecking up feed will have no particular effect on their growth. In that savage society of eighteenth-century France, fashionable ladies kept various ornamental birds which had half of their upper beak removed. This made it completely impossible for them to feed themselves, and they became totally dependent on being hand fed each bite. This dependency made them most tame and servile to their mistresses. But debeaking as it is practiced in poultry husbandry is painless, or nearly so, and it leaves them with no such debility.

Cannibalism is not restricted to chicks. However, adult birds in small flocks are not as susceptible to cannibalism as chickens that are crowded into large flocks. Sometimes in adult flocks cannibalism may reflect a dietary deficiency. Preparations of mineral additives are commercially available and are administered through the drinking water, which may have value as control or prevention. Consult your county agent, or ask the manager of your feed store (or druggist, if you live in a rural area), should the problem arise. Adult birds as well as chicks that have been the victims of pecking need to be treated with medication for purposes of healing and to deter further cannibalism. Antipecking medication contains a dye of a particular hue and intensity that is unattractive or repulsive to chickens and helps thwart further attacks.

Laying hens are occasionally unable to pass an egg through the vent. In this condition, the hen is said to have an egg "hung up." If the egg is visible, you may try breaking it with the point of a knife, etc., and release the passageway. It is unlikely that such a hen if she recovers will ever again be an egg producer. If the egg is not visible, you will detect a hen with a hung-up egg by her ungainly and uncomfortable gait. Often she will squat on the floor or in the yard and hardly move at all. The vent will be somewhat distended and show a discharge. Sometimes your first clue to a hen being in this condition is the other chickens pecking at her bottom. If the hen is in this condition, by all means dispatch her promptly.

Besides being prey to themselves, chickens are subject to depredations from other species with a more purposeful appetite. Rats may and do come into a brooder house and feast upon young chickens, although they more rarely attack adult chickens. Rats also consume enormous amounts of chicken feed. The most effective control is prevention. A properly built floor can prevent them from gaining access to your chicks. It is essential in control of rats to eliminate places that might harbor them. If you maintain a droppings pit under the chicken roosts, this can provide rats with a natural habitat if it is not cleaned out periodically. The accumulation of rubbish, piles of lumber, bales of straw, etc., near your chicken house or pen provides refuge and breeding grounds. The rats themselves are not difficult to kill. Unfortunately, unless you can find a burrow or run that is inaccessible to the chickens themselves—or pets—it is a problem to trap them safely. If it is feasible to trap, the best system is to use a spring trap, inserted far enough inside a burrow or run to make it inaccessible to chickens or pets. No bait is necessary; you hope, with reasonable confidence, that the rat will inadvertently put a foot in the trap as he travels by. Poisoning is possible, but by all means check with a county agent to select a poison that will be safe to use, and use it exactly as directed.

In some areas weasels may also be a source of trouble. They may handily kill much larger chickens than rats can manage and are fond of eating eggs as well. Like rats, their activities will usually be nocturnal. Fortunately weasel infestations are usually limited to one, or a single pair. They are difficult to trap unless you can find their run, and very difficult to poison safely. The best defense is prevention—a tightly secured chicken house. If you are troubled by a weasel and can find a hole he travels through, because of his long slender body he may best be caught by using a gopher trap rather than a spring trap, and the construction of this type of trap makes the likelihood of harming a chicken or pet unlikely. Unfortunately this type of trap is not very successful with rats.

Skunks are lovely things, but an abomination among poultry. Because of their natural defense, they have a little too much self-confidence for their own good, but happily this helps make them easy to control. They eat baby chickens, up to a fairly mature size, and are mad about eggs. If nothing is done to control them, they will establish regular skunk condominiums in your neighborhood and rear whole generations on your eggs and chicks. Their fearlessness makes them easy to trap, but their fearsomeness makes them dreadful to catch. If trapping is resorted to, especial care must be taken to prevent your pets getting into the traps. The best solution is to set the traps at night after locking up all your pets, and spring the traps in the morning before releasing the pets. I have found an effective system to be to set up a dummy nest outside the chicken pen, baited with a few eggs, with two traps set inside.

I have had particular trouble with skunks raiding duck and goose nests. I have found that invariably if they raid the nest one night, they will raid it the next. Sometimes the birds themselves will give the alarm when the skunk shows up. Otherwise, periodically checking with a flashlight eventually turns up the raider in the nest or nearby. Caught with egg on their chin, they will usually give you an impudent stare and go on eating, daring you to do anything. The thing I do is to shoot them. I realize that this system of control runs counter to some people's sensibilities, and I respect their reservations or downright rejection of the idea. Like many people, however, I grew up considering a gun to be a quite normal and useful household implement, like a shoehorn or a runcible spoon. And while I respect the humanitarian attitude that would reject killing the skunk, I also respect the goose and her nest. After all, she's got a lot invested in the nest herself. Ultimately, if you can't resolve your attitudes toward the question of predator control, the balance of nature may well take the amateur production of eggs and chickens out of your hands. It's a kind of heat you'll have to stand if you're not going to be forced out of the kitchen.

With chickens running at large in a pen, hawks may occasionally create a predator problem, particularly in remote areas. For the most part hawks prey only on the younger birds. Probably the best solution, if hawks begin to present a problem, is to confine the chicks until they are large enough to no longer be especially attractive. The presence of a rooster is a distinct deterrent to attacks by hawks. Instinctively, the rooster lets out a particular shrill warning cry when he sees a large bird overhead (or anything else: a kite will throw a rooster into paroxysms). The hens respond by taking evasive maneuvers, and a hawk can seldom succeed in capturing an agitated hen. In fact, some roosters seem to develop a technique for obtaining a sort of unearned prestige by being hawk alarmists. Endlessly through the day they will periodically give out with the hawk alarm and watch with arrogant satisfaction as the hens fluster and run for cover. But unlike in the story of the little boy who cried "Wolf!," the hens never seem to become unheeding of their rooster's warnings. I suspect the fact that the flock survives each attack gives prima-facie evidence to the hens of the rooster's competence.

In the past, in remote farm areas, I have known hawks to become serious pests. Usually it would turn out that a single hawk was responsible for all of the losses. Elimination of that one hawk would

Chickens Cantankerous and Droopy

usually take care of the problem. Many or most hawks are now under protection, so if you choose direct action, check the law before you move. Personally I am so fond of hawks in general I would be hard pressed to kill one even if I caught him red-handed.

In different parts of the country, certain other birds may prey on baby chicks or steal eggs. Here in the western states, magpies often raid nests, particularly the nests of such breeds as bantams or Guineas that hide their nests and raise their broods in the fields or cover. Magpies will sometimes enter a chicken house, but generally they are too wary to take that risk. I have found that shooting one or two magpies, even when they are numerous, will make your estate cease to be a hunting ground for the survivors. The brains that have been bred out of turkeys seem to be concentrated in the magpies.

I have never had experience with foxes, but their western counterpart, the coyote, can be a real menace to all kinds of poultry. The best defense against coyotes is a good dog. He will not kill them, unless he is an exceptional dog, but his presence will keep them from coming in and pulling off a raid. Coyotes also have a particular sweet tooth for cats and may even kill off the cats before they begin investigating the poultry.

Probably the most troublesome predators of all are domestic dogs and cats. Your own dog can easily be trained not to molest your poultry. More often than not, he will actually come to enjoy them, as dogs so often become solicitous of things belonging to their owners. It is not unusual to see a dog which is very protective of the poultry, and I have seen dogs trained to herd turkeys in the same way that sheepdogs herd sheep. The problems dogs present almost always are problems with strays, or dogs belonging to a neighbor. An amicable relationship is indispensable in working out this kind of problem. But unless you have a very strong fence, ultimately everyone will face the grief of finding a dog or dogs have broken in and killed a large number of his birds—chickens, ducks, or whatever. There is probably legal recourse if you care to pursue it. The code of the West has it that any animal which molests any livestock has forfeited his niche in the ecocycle.

Cats are a real hardship to small chicks. Especially the passing stray or curious wanderer may make irregular or regular raids on one's flock. Usually they will not initiate wholesale slaughter as a dog will, but a stray may make repeated raids, picking up a tidbit every night or so, and eventually reckon up a serious score. A family dog is a good deterrent to trouble with stray cats. Happily, one's own cats usually learn, in mysterious ways, to leave the chicks alone. The half-grown cat, or the female with her first litter of kittens, is the most likely to cause problems. I always explain to our cats that if I discover them catching baby chickens I will donate them to the National Bureau of Standards, where they will live out their lives helping to determine the exact size of rooms not big enough to swing a cat in, and bringing the standard of scientific precision to the speed of a scalded cat. The lectures seem to work.

Housing the Humble Chicken

Probably the simplest shelter that exists, short of a cave or a thicket of bushes, has been, in the vocabulary of tradition, a chicken house. Something in the word itself conjures up the idea that a chicken house is intrinsically, or even necessarily, ramshackle. Like fine old New England families and their money, most established farms have their chicken house. If you must make your chicken house, rather than inherit it, consider it a good opportunity to try out your architectural ingenuity and an opportunity to fumble through the skills of basic construction. It doesn't have to be a palace, and the hobbyist's chicken house is often or usually made from salvaged materials or the cheapest grades of building supplies available.

It is not practical to give specific designs, or provide actual plans to follow or choose from, because the variable of specific operations are too great. These variables are what give chicken houses their unique character and distinctiveness. What is required for an adequate chicken house in one part of the country will be quite different from what is preferable in another part of the country. Agricultural colleges in the different states and regions have extensively studied housing requirements and have published their recommendations complete with plans. Ask your county extension agent; he can undoubtedly put you onto government publications directed toward the requirements in your region, with plans for a variety of structures designed with exactly your requirements in mind. While much of the research of the agricultural colleges has been conducted for the benefit of commercial operations, the needs and requirements of the small-time operator have not been neglected. Besides your geography, the kind and size of your poultry enterprise will be major factors to consider in selecting a design. If you are raising broilers and fryers during the warm months, even in the northern states it will be necessary to make only small provisions for cold weather, and lighting requirements will be entirely different from an arrangement made to provide for laying hens. If you are raising only layers, or combine raising layers and fryers, specific requirements in addition to fryer facilities must be considered.

Assuming both operations, egg and meat production, to be small of course simplifies the planning. It may be possible to make use of existing facilities, or to modify facilities which were originally intended for quite different uses. Probably the commonest facility modified is a garage, or part of a garage. Having found keeping more than one car economically unsound, many people have modified half of their two-car garage into chicken facilities. Sometimes a shed and even an outgrown childrens' playhouse have been modified. Many small prefabricated structures intended for storage, or designed originally for lawn and garden equipment, may be available, and possible to modify. Prefabricated structures may be

Housing the Humble Chicken

purchased new, but probably the cost would be greater than building a structure of equal size from salvage or cheap grades of building supplies.

Ease of cleaning is a major factor to consider when planning a chicken house. Some chicken houses are built simply using the ground for a floor. This type of construction is cheap and simple, but not the easiest to clean. It entails a generous use of litter and invites infestation of vermin, particularly worms. Sometimes a dirt floor with a sand or gravel surface is used to help combat worm infestations, but cleaning entails frequent replacement of the filler material. Where temperature control is no problem, sometimes the

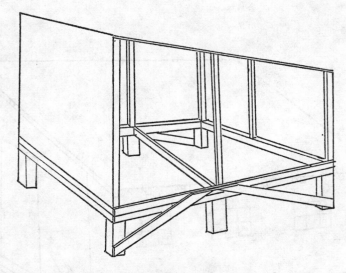

In climates which have little cold weather, a mostly open shelter of this design is adequate. The sides and floor are chicken wire. Nests, feeder, and waterer are attached to the solid wall and are serviced from the outside through access doors. Pole perches or slat roosts may be provided.

chicken house is raised and a slat floor, much like the "duckboards" in barracks and gymnasium showers, installed. Droppings fall through the slats down to the ground below and are removed from the outside by raking. Wooden floors, used with litter, are common, as are wooden floors covered with some of the tougher synthetic surfaces now available. I floored my present chicken house with ¾-inch plyboard, ½-inch particle board, and an upper surface of Formica board. In this colder climate that surface provides a good insulation and has proven to be easy to clean. All materials were utility grade, so the expense was not staggering. Large operations frequently use concrete floors, which are wormproof, and the surface may be easily cleaned with running water and brushes.

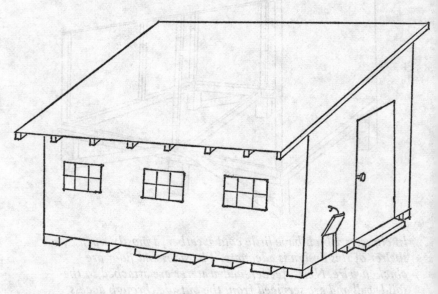

My chicken house.

Housing the Humble Chicken

The frame of the house should be weatherproof. Probably cost will finally determine the material used. Unless scrap lumber is being salvaged, the cheapest material for the outer shell and roof will be plywood, weatherproofed with tar paper, roofing paper, roofing tiles, etc. Small lots of roofing material may frequently be purchased very cheaply, and most chicken houses require only a small lot.

Aesthetically, a chicken house constructed with the main eye to economy will be a little dowdy in appearance and perhaps ostentatiously utilitarian. A little deviation from basic practicality can smarten up a chicken house considerably. Try dressing it up with a shingled exterior. For a small house, the cost will not be out of reason, and its heat retention will surely be improved. Or try some structural innovation; perhaps you can improvise on a basic design which is starkly utilitarian, making the chicken house match or coordinate with your own home, or frame it to look like a miniature barn, chalet, etc. Remember, it doesn't have to *look* like a chicken house to do the business it's intended for.

The interior of the house, however, will be almost exclusively utilitarian. If you can justify the expense, running water is useful, but with a small flock formal and permanent plumbing is almost never economically practical. For a slight expense, however—the cost of a length of garden hose or plastic pipe—a permanent-temporary arrangement can easily be built in to the chicken house, connected with an outside standpipe or sill cock only as needed. This arrangement can be made with absolutely no knowledge of plumbing and no disruption of your regular plumbing system. The only technical requirement is that the outlet be structured higher than the intake so the hose or pipe can drain completely and prevent freezing in winter.

Electricity in a chicken house is not indispensable, but is certainly not merely a luxurious laborsaving device either. If you intend to brood baby chicks in it, it will almost certainly be a necessity. For winter heating, if you intend to supply it, electricity is usually the only practical source of energy. If hens are to be kept laying in the winter, they will need to be supplied with both heat and artificial light. Therefore even a small chicken house should make provisions for at least one light, plus an electrical outlet. Wiring a chicken house need not be expensive, but it should go without saying that to install it you should either know what you're doing or enlist the service of an electrician. Unlike with water, temporary-permanent arrangements are likely to be hazardous to life and property.

The chicken house in which laying hens are to be kept will most likely have roosts, although commercial operations raising hens unconfined frequently do not. The old-fashioned sloping-pole roosts have been replaced by superior systems. For the small or medium-sized chicken house, probably the best arrangement is with the so-called "droppings pit." It really amounts to a sort of crib about eighteen inches wide, raised about eighteen inches above the floor.

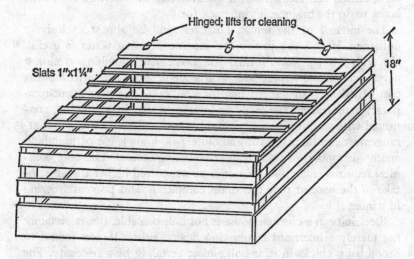

Droppings pit. Length and width depend upon the size of the chicken house and the number of chickens.

The top consists of slats 1¼ inches wide and 1¼ inches apart. One or both of the two ends will probably be positioned against the walls of the chicken house. The back will be against a wall of the chicken house and the front should be solid, made from either planks or plywood. It is possible to design the pit so that either the front or the top is easily removable. A common design is to hinge the top to swing backward for easy cleaning of the droppings. The hens will roost as they please on the slats, and as the droppings accumulate they are removed; probably with a few laying hens it will only be necessary to clean it two or three times a year. Practically, it will be found that the hens will spend a good part of their waking period on these roosts when they are not outside, which further simplifies cleaning because the droppings of the day tend to accumulate in that one spot. A sheet of chicken wire should be attached immediately under the slats. Any carelessly laid eggs will be caught and easily retrieved.

Nests should be provided on a ratio of at least one nest per every four hens. If you are involved with testing and breeding and using trap nests, ideally one nest per bird should be allowed. Conventional nests should be sixteen inches deep and fourteen inches on each side. Wooden boxes, such as apple boxes and orange crates, frequently are used, but they are a little cramped. Leghorns easily use these smaller nests, but the larger breeds are uncomfortably cramped by the size of most boxes. If the nest is at all cramped, eggs will frequently be broken. It is desirable that nests be filled with straw, wood shavings, excelsior, etc. Such materials both protect the eggs from breakage and help keep them clean. Nests which are allowed to become fouled will usually result in stained eggs.

Feeding arrangements in the chicken house should be convenient for both the chickens and the farmer. Feeders which are hung from the ceiling may hold enough feed for several days at a time. A suspended feeder helps reduce the amount of feed the chickens scatter and prevents the chickens fouling the feed. "Fly up" feeder bins may also be used. Like the suspended feeders, the feed is gravity fed into the feeding trough. The chickens must fly up a foot or so and rest on a perch to eat. Depending on the size of the flock, a feeder of this sort needs to be filled only once a week or so. It is possible to design such a feeder into the chicken house when it is being built, to allow filling from the outside.

If no running water is available inside the chicken house, a gravity-fed waterer is almost a necessity. Made of galvanized metal, they come in rather large capacities. Like feeders, some are designed to be hung. Despite all precautions, they will spill over a little; therefore, they should be positioned so that accumulation of moisture presents no problem, and where litter, etc., may help to absorb

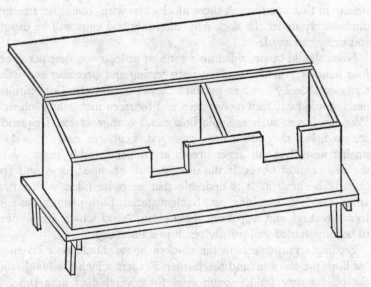

This nest has a hinged door, which keeps the interior dark. Hens enter through the two ports. Since hens will leave the dark interior as soon as they lay, there tends to be less breakage or soiling of eggs.

Housing the Humble Chicken

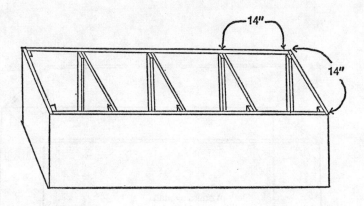

This common, simple type of nest may be either attached to a wall or given a stand. A rule of thumb is to provide one nest for each four hens.

the moisture. In warm weather, portable waterers may be moved outside. In cold weather, unless the chicken house is heated, steps must be taken to prevent the water from freezing. Not only does freezing prevent the chickens from having their drinking water, but also when the water freezes it will rupture the waterer. If you have an electric outlet available, immersion heaters are available which will keep the water at a controlled temperature. The heater for a tropical-fish tank works fine.

If space limitations are too restrictive to allow an actual outdoors run, a sun porch may be designed to provide at least limited access to sunlight and moving air. If a sun porch seems desirable, your particular situation will largely determine design and size. Usually a sun porch is attached to the chicken house, elevated some eighteen inches above the ground. Up to a point, the size will be determined by the space you have available. Construction is simple, since it will

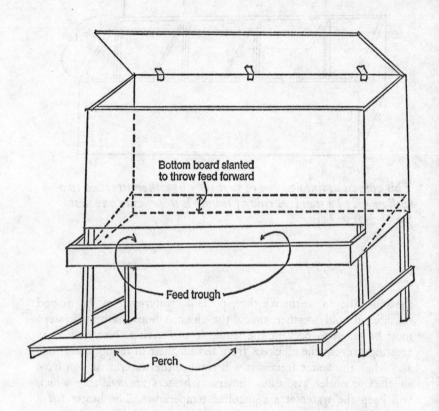

Freestanding fly-up feeder. They may also be attached to a wall and built to any convenient size. They may be easily fabricated of ¾" plywood on a 2×4 frame. The perch may also be made of slats.

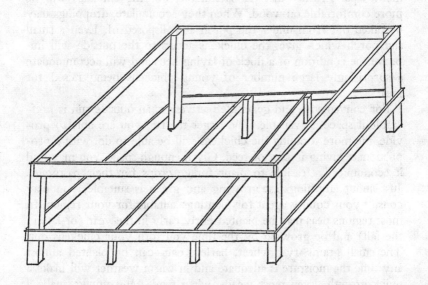

Where space is limited, a sun porch may be attached to the chicken house to provide sunshine and fresh air. The frame is completely netted with chicken wire. A stronger floor can be constructed of steel netting or wooden slats. The dimensions of the frame may vary from 5'×6' to 12'×12'.

be, in effect, no more than an elevated chicken yard. To provide afternoon shade, the best location is on the eastern side of the chicken house. It should require no roof, since, in inclement weather, birds may take refuge inside the chicken house. If an eastern situation is not possible, some kind of sunshade might be required, at least for the warmer climates. For ease in cleaning, the floor may be made of welded steel mesh, or of wooden slats some 1⅛ inches wide and 1⅛ inches apart. Either material is suitable, but the chickens will be more comfortable on wood. When they accumulate, droppings may be raked out from under the porch and disposed of. Even a small sun porch which gives the chickens access to the outside will improve the condition of a flock of laying hens and will accommodate a surprisingly large number of young chickens being raised for meat.

For convenience and general effectiveness, an outside run is preferable if space is available. The larger the run you are able to provide, the more foraging the chickens will be able to do, with a proportionate saving in cost of feed. Given enough space, you may find it economically feasible to plant forage crops for their pasturage. Just about anything fast growing and green is suitable; you may consult your county agent for plantings suitable for your region. In most regions peas may be planted fairly early in the year (or late in the fall) and be growing by the time you turn your chickens out. The small grains—rye, wheat, barley, oats—can be planted almost any time the moisture is adequate and in warm weather will make a quick growth. Even most weeds, when tender and young, make a forage that the chickens will eat.

The breed of the chicken will be the major consideration in determining the size and construction of an outside run. Leghorns have small bodies and an advanced talent for flying. The standard small-meshed, uniform chicken wire is the only thing that will contain them. For the larger breeds of birds, the coarser square or oblong graduated mesh—small at the bottom to larger at the top—is suitable. Although the standard-mesh chicken wire is cheaper, the larger mesh is longer lasting and somewhat easier to install. The quickest and easiest method of erecting the fence is by using steel fence posts. Steel posts have the further advantage of versatility; they may be easily uprooted and moved, making it possible to change the size, shape, and arrangement of the chicken run as circumstances require.

The flying habits of Leghorns makes them an especially difficult breed to contain. Old-fashioned wooden fence posts, with a fourteen-inch one by four nailed on the top of each post, cantilevered toward the inside, are the classic country construction for containing them. The chicken wire is attached all the way up the post, and a fold spread up on the top of the cantilever supports facing into the pen. To make the fence completely secure, the bottom must be fixed to the ground. I have found that short, hooked sections of stiff wire (pieces cut from coat hangers are ideal) driven into the ground about every sixteen inches hold the wire firmly and permanently in place. Such a fence is virtually chickenproof and will even contain the adventuresome bantam. It will not contain Guineas, however. Only a complete overhead mesh would cut them off from flying.

However you design your fence, remember that while its primary purpose is to keep chickens in, it should also keep unwanted invaders out. It will probably not be ratproof or weaselproof, but it will be a deterrent. It should strongly deter skunks, cats, and dogs except the most persistent and bloodthirsty sort.

The Feasible Chicken
≪≪≪≪≪≪≪≪≪≪≪≪≪≪≪≪≪≪≪≪≪≪≪≪≪≪≪

For the small-time hobbyist, the balance sheet is more likely to be an embarrassment than a subject for bragging. Most who attempt to put their marginal operation on a paying basis by selling surpluses—eggs, fryers, roasters, etc.—are apt to find, when they reckon up their accounts at year's end, a strong and pungent taste of shoestring in their mouths. Good accountants probably do not make contented amateur chicken ranchers.

By raising our own fryers and roasters, I think we have shown a distinct savings in money over the retail price of the same quantity of meat. For our family's use, I generally plan to raise about fifty a year. Of these, the majority are killed as fryers, and frozen. Usually I keep about ten to mature for roasting chickens. To keep things in proportion, fifty fryers dressed out at four pounds each provide two hundred pounds of meat; at five pounds, two hundred ten. Raising Hubbards, the roasters often dress out at ten pounds each. Making the wild assumption that you raise fifty of them to this weight, that would be five hundred pounds of meat, equivalent to the dressed weight of a one-thousand pound steer. In other words, it is clearly true that you can raise one heck of a lot of meat in a very small space. Let me hasten to explain that while I like chicken, the thought of eating five hundred pounds of it a year gives my stomach a feeling that is other than wholesome.

Costs fluctuate so widely—always in an upward direction, it seems—that any specific dollar estimate of the cost per pound of meat would be meaningless. As a general rule of thumb, our home-grown fryers seem to cost less than half the retail cost of dressed chickens. If your circumstances are right, you may do a bit better than that. Or worse. The longer you keep the birds after it is possible to slaughter them as fryers, the higher the investment in feed and, ultimately, the less "profit" per pound of meat.

The real black ink in your ledger book is not monetary, however. At ten or twelve weeks of age, your chicks should be frying size. Short of catastrophe, you are almost certain to at least break even on feed costs versus retail price of chickens of similar weight. What the balance sheet will not reveal is quality. Home grown *is* better, and it tastes the way chicken used to. It is so much better that a whole new standard of judgment is required if you have never had it before. Even our children, whose conditioning by advertising, peer pressure, or whatever makes them prefer bakery-shop sweets and Styrofoam cupcakes to the products of their mother's kitchen, have such a preference for our own chickens to those commercially obtained that they will hardly eat any other. If you want chicken to taste that good, you will simply have to grow it yourself. It is this superior eating quality that makes it desirable to grow your own even if you do not make an appreciable financial savings.

The Feasible Chicken

The economic advantages of home-grown eggs is somewhat more difficult to document than the comparatively streamlined and clear-cut operation of raising fryers. It is best to consider your poultry operation in its entirety before beginning a ledger sheet. If the physical plant is amortized in two directions, both meat and eggs, you are more likely to justify the economics of a small-scale egg production. For example, $100 will buy an awful lot of eggs; it will not buy an awful lot of facilities and feed. If the facilities produce fifty fryers and then double the rest of the year as the habitation of a family-sized flock of laying hens, overhead may be spread and partially absorbed. If you wish to build facilities for six hens and maintain them, it will be a very long time before you break even financially.

It is healthful economics to consider the initial investment from a different direction; small-scale poultry raising is a *hobby*, and part of the expense should be written off as being in the interest of pleasure. For example, many of us maintain a tank of fish. The cost of facilities, plus fish and maintenance, makes a reasonable analogy to the cost of housing and maintaining half a dozen hens. Many tropical fish cost the equivalent of a started pullet, a few the cost of a whole penful. Or consider other domestic pets. Our household cats eat more in actual food cost than our flock of laying hens. The accumulated vet bills of our patriarchal tomcat equal the cost of all the poultry furniture I own.

As with any other pet, I feel I can justify the laying hens I maintain purely for the pleasure keeping them gives me. I like to tend them, watch them, and especially I like to gather their eggs. The facilities are there; their upkeep is less than the cost of our cats. And the *chickens* give us superlative eggs. When they are laying heavily, our half dozen or dozen chickens produce vastly more eggs than our family can consume. Strictly on the basis of feed consumption, when they are producing well the eggs are much cheaper than eggs commercially obtained. Like the meat of home-grown fryers, the eggs of a home flock are superior in taste to the eggs of commerce. The surplus goes to friends and yields a return in the pleasure of giving and, in addition, often returns in kind from other people's home produce. The ledger would say that the egg end of my poultry operation is a losing proposition; I know otherwise. A balance sheet is mute to the intangible profits of hobbies.

Most growers of home poultry will develop their own systems and techniques of cutting corners in the cost of producing both eggs and meat. I have mentioned the possibilities and advantages of using household and garden scraps, scrounging from produce markets, and in a limited way growing your own feed. You should also consider the manure as a byproduct not to be despised and also a product that may well replace or certainly supplement an actual cash outlay for fertilizer. Among animal manures it is the richest in nitrogen, although lacking in phosphorus and potash. It may seem an invasion of privacy to say it, but each hen produces about fifty pounds of manure a year. A recommended rate of application is about one bushel of manure per hundred square feet of ground. For balance, it should be supplemented with about a quart measure of 3-8-7 prepared fertilizer. Or better, the chicken manure may be composted with grass clippings, vegetable wastes unsuitable for chicken feed, and other organic wastes and applied without supplement.

During the fall and winter, the manure may be applied to the land as it is removed from the chicken house. It is an elegant lawn fertilizer when applied during the winter and can be applied quite lavishly. Because of its high nitrogen content, when it is applied during the growing season in too great a quantity it can cause "burning" of plants or grass. Rather than applying it during the spring and summer, the manure is best handled in a compost heap or pit. Since its decay helps break down other organic material quickly, it is most efficiently used by putting down alternately a layer of refuse, a layer of manure, a layer of dirt, always ending up with a layer of dirt on the outside. If it is not composted, and piled outside, it should be covered with a layer of dirt, a sheet of plastic, etc., or rain water will leach out much of its nutritional value. Because of its high nitrogen content, I always reserve the poultry manure for our roses, our salad garden, and our strawberries. If you're still accounting, the supplement to, or replacement of, fertilizer which must otherwise be purchased adds a real and tangible smear of black ink into the register. You don't get that advantage with tropical fish.

For the hobbyist, ultimately the accounting turns into a kind of habitual and closed-circuited system of rationalization. The hobby is its own justification, and the only reason it seems to require apologizing for itself more than other varieties of pets is that unlike most pets, chickens have a clear utilitarian feature, which therefore suggests utilitarian standards must be applied in justifying them. Since it is utilitarian, reason tells us it should be paying. But one must quickly add to the economic utilitarian features the quality features as well. A garden, for example, clearly has economic value, but probably most gardeners are more attracted by the quality of the produce than direct economic savings. But if the gardener didn't like doing it, quality or no he probably wouldn't.

The pleasure of keeping poultry has no more need of justification than raising a garden. Indeed, over the course of time I have observed that everyone who tries to defend objectively what he does for pleasure—fishing, hunting, bird-watching, gardening, whatever—ends up making a fool of himself. One takes pleasure in the things one chooses to do because the human being seems unique in creation in being able to take pleasure in doing nonutilitarian things, or things with only vestigial utilitarian value. It takes a little human bravery, and enhances one's human dignity, to say, "I do it because it is my whim."

When a Sparrow Falls

People who raise poultry on a small scale, or who look with favor on that sort of avocation and read books on the subject, most likely have affection for birds of all species. Affection and attachment to birds other than obviously useful poultry is sort of curious in a way; most human/animal bonds have at least some sort of vestigial or faintly historical usefulness behind them. In the case of birds other than poultry (with the single rare exception of hawks), the reason for the bond is simplified to the point that only natural and unselfish affection can account for it.

Those whose affection for birds is exceptionally strong almost without exception eventually will find that they have taken on the care of an orphaned nestling of some species, usually under circumstances of most awkward and inconvenient timing. The immediate questions are what to feed it and how generally to care for it, and there is usually no reference and no expert to consult. If the nestling is extremely small, unfeathered or completely naked, the odds against success are frankly staggering. What to feed it depends on what it is, and a good guess at what it was being fed in the nest usually is not difficult. In a pinch, the yolk of a hard-boiled egg will be accepted by most small birds, and it seems to agree with them particularly if supplemented with "natural" foods. Even if known, the natural foods are simply not that easy to come by in an emergency. Also bread crumbs soaked in milk are often accepted by nestlings and can be used as a nutritional substitute, if not a main item of diet. Even if the nestling is well feathered, it must be provided warmth; a heating pad or light bulb is usually the most convenient source of heat for a homemade nest. If a nestling is taken on to be raised, the ultimate goal always should be to raise it in such a way that it may be released when mature and be able to survive as a wild bird; and often the final return of the bird to the wild is the most difficult stage of the "confinement."

It is unnecessary to remind those with genuine affection for birds that a nestling must never be pilfered. It is nevertheless true that, in the course of quite natural events, nestlings do fall by the wayside, and some invisible law always brings them to the doorsteps of people who care. An immediate word of caution is in order, however: In the ancient legal formula of "Did he fall or was he pushed?," in the milieu of birds fallen from nests, the answer most usually is, "He was pushed." As tender as birds are to their young, it seems that given a choice, ultimately their loyalty is to the species rather than to individuals. Most often it appears that when a hatchling is defective, it is pushed from the nest, "culled" as it were, by its own parents or more likely by its siblings. While it is certainly possible that a baby bird found on the ground simply fell out or was tipped out by heavy wind, or even fell because of a poorly engineered nest

When a Sparrow Falls

(some birds never seem to get the hang of nest building), remember that baby birds are professionals at the art of staying in their nest until they are mature enough to survive outside of it. It is fairly likely, therefore, that a bird found on the ground is defective. Birth defects, disease, infestations by parasites, such things as would make it unlikely to survive or which could even sap the genetic strength of the species (such as having parents which build poor nests!) most often result in the infant bird being rejected. Such defective birds may sometimes be hand-raised successfully (I can think of at least two books written about such successes), but the chances are slender that the bird could ever survive in the wilds. A summer ago my eldest daughter brought in two foundlings a week apart—an English sparrow and a finch—and both lived only a few hours. The finch was utterly and horribly infested with worms in its flesh, and the sparrow had some undiagnosable but quickly mortal internal disorder.

The usual healthy orphaned nestling one encounters has been unnested by a storm, death of the parents, or by children. The first orphan I ever raised successfully was a baby owl—a western field owl—that I suspect had been "liberated" by a boy and his father. At the time I was only eight or ten years old and thought it was good the chick had been rescued by someone with the foresight to bring him to me.

The owlet was a sort of nursery-book kind of creature, looking nothing like an owl, a bird, or anything else of animal kindred. For all the world it most resembled a ball of white angora knitting yarn, about the size and shape of a goose egg, and which persisted in defying gravity by resting on the smallest end. Only the closest of inspection could locate eyes inside the white yarn, and on discovering them, no matter what direction you "faced" it, the eyes were always fixed steadily on you, although the ball of yarn had never moved. The phenomenon was slightly preternatural. Likewise, in no matter what spot your finger touched the yarn, it was instantly bitten. Sharply and painfully.

I had not the remotest idea of how to go about caring for a

young owl, and only by dumb luck and chance did I do the right things. There was only one food available for me to feed him—sparrows—and fortunately there were lots of them. For the first few days I "dressed" them for him, but as he began rapidly to develop feathers and shed his yarn-fur, it was clear that the owlet could, and preferred, to take care of that bloody business himself. His eyes became balefully visible and his beak appeared, second in prominence only to his eyes. And he grew.

When it was given to me, his abductors said they thought it was a ground owl—a familiar inhabitant of the untilled sagebrush hills. They often inhabit abandoned badger holes and coyote dens, and a good-sized colony of them may present you with the spectacle of more than a dozen standing soberly around the trampled mound of dirt at the entrance to their burrow. Always, all eyes will be focused on an intruder; no matter how you turn, the eyes always follow, but the head never seems to reverse direction. Finding one perched on a fence post, I had tried the macabre and juvenile experiment of slowly circling the post to see if as the head persisted in following me in a continuous circle the bird would eventually wring its own neck. I really knew it wouldn't happen, but no matter how closely you watch, your eyes are never quick enough to see the moment when they reverse their heads. For all intents and purposes they will rotate their heads endlessly in whatever direction you choose to pace.

The ground owl, or burrowing owl, is quite small, only little larger than a robin. This owl, however, grew and throve until it became obvious he was definitely an owl of another species. It was finally determined that he was a species locally called simply a "field owl," a medium-sized owl that was common in the same area the ground owl lived. They often hunt by day and learn to follow farm machinery around to catch the rodents the equipment scares up or turns up. While harvesting wheat, I have seen them, together with hawks, in flocks of perhaps fifty, following the combine in great circles, catching mice. Not infrequently a careless or stubborn owl would not move quickly enough and would find it had in its turn been ingested into the combine and threshed.

But regardless of species, it was my owl, and I was stuck with him, or we were stuck with each other. Every day required I provide him food, and his appetite was enormous. Equipped with my Daisy single-shot BB gun, my daily chore was to hunt for him. I also set traps in the barn and feed sheds for mice, which helped take some of the strain off me when other duties pressed. When the owl was fully grown, and school had opened in the fall, it was very helpful on occasion to be able to fall back on the yield from a successful trap line. The daily responsibility for sparrows did wonders for sharpening my marksmanship, however.

Our house had a screened porch, which became the owl's semiprivate home. For some reason, even as an adult, he seemed to like to retreat into the box in which he had lived as an infant and in which I continued to feed him. A window of my bedroom looked out across the porch, and often in the night I was awakened by his fluttering back and forth in the dark. Even though I knew the source of that muffled, eerie sound, being a small boy, in the blackest hours of the night I would convince myself that the porch was haunted.

Never in our relationship did we truly become friends, or even really easy with each other. When I offered him a sparrow, he would usually remain absolutely motionless for some time, then snatch the small carcass from my fingers in his talons with such a speed that I could hardly see what happened. If he ever showed me any affection or consideration, it was only in not biting me quite as hard as he did other rare and incautious visitors. He seldom drew blood from me; others, he invariably made bleed. It did not strike me as strange or unusual at the time, but I have since raised many small birds of widely varying species, and invariably they would "imprint," or identify with me, and become companionable if not downright affectionate toward me.

As the winter wore on into early spring, he became more and more restive at night, and the swoops across the porch, just a feather's whistle short of totally silent, became more and more disquieting to my sleep. Finally I decided the season was advanced enough, and he was old enough, that he could survive on his own. After a solid feeding consisting of several sparrows and a good plump mouse, just at dusk I set him and his box outside, leaving the screen door to the porch ajar as a retreat in case he lost his nerve. I watched until complete dark had settled, but he did not move; however, my sleep that night was untroubled, and the next morning he was gone.

For perhaps a month he more or less maintained a headquarters around our house, although I never saw him again. He quickly took up a practice he had never indulged in captivity, and hooted, a soft, almost seductive *coooo*. For the first few days after his release I made sure there was fresh food in his box, but he never took it. Sometime in the spring he left for good. I hope he discovered about the good and easy hunting behind the farm machinery.

My second experience raising a bird of prey was on a soberingly larger scale and many years later. Unexpectedly, and without preparation, I became the protector of an orphaned baby golden eagle. The success in raising the owl had depended mostly on luck, but with the eagle, I read as much as I could on all birds of prey as well as eagles and went about the whole thing scientifically. Through my reading I discovered with the owl I had just happened to do exactly the right thing with diet—feeding whole small birds and mice. The digestive system of birds of prey is designed to turn the seeming inconvenience presented by fur and feathers into an advantage and necessity.

That the eagle came into my hands at all was a by-blow of fate. It had been orphaned in eastern Colorado by the legal killing of its parents on a writ of lamb killing. In this case it could be said that technically I raided the nest, but it was a nest inhabited by a doomed orphan. If it had been a dangerously placed nest, I would not have made the raid, but, unlike the eagle's nests of romance which are located among isolated pinnacles and stone cliffs, this one was situated mundanely on a steep but ordinary dirt bank, in the central Colorado "outback."

My reading had suggested that when there was a lone eagle in the nest, the chances were long that it was a female. Being stronger and more aggressive, the female chicks tend to chuck the relatively more phlegmatic males over the side. In every respect but size, this was a typical baby bird—semihelpless, covered with down, appealing—but the size of a small Thanksgiving turkey. Its huge feet seemed nearly adult-sized, but the bird was helpless to stand erect on them. Its head was mature, even massive, but its beak capable only of feeble pinches. All of its instincts seemed to yearn to unite the efforts of the beak and talon, through the solid train of muscles between, for ripping and tearing.

In the case of the owl, its food had been simply what a country boy could provide—sparrows and mice. The diet of the eagle would have to be provided for by planning. Books suggested feeding animal organs for the main diet of most birds of prey, and with a little experimentation, taking into account both her taste and the tariff at the meat counter, kidney became the staple of her diet. I knew by this time that all birds of prey required the feathers and fur chance had provided for the diet of the owl. Under the present circumstances, however, a substitute would be required, and after a little experimenting the most suitable and convenient substitute was found by adding thread raveled from old white dress shirts. A mucous byproduct of digestion accumulates in the craw of the bird, which should not pass through the lower digestive tract. Nature had provided that this mucous be absorbed by fur and feathers in the upper end of the tract and expelled through the mouth. This process is called *casting*, and the pellet of fur and feathers (or, in this case, thread) is called a "casting."

In addition to nourishing food, a bird of prey also requires what might well be called "digestive exercise." That is, the bird requires not only food as nourishment but it needs as well the psychological and physical exercise of tearing up meat. This exercise of ripping meat from bones is called *tiring*, and the object itself being worked over is called a "tiring." For the eagle, I provided chicken backs and wings, which were both inexpensive and satisfying to her. Tiring on them, she gave the same reflective sense of a dog quietly chewing a bone in the shade, or a cow relaxing in the grass ruminating with her cud.

The size and primitive power of an eagle makes it an awesome, even intimidating bird. With quiet but not excessive handling, this one seemed to adapt easily to living with me and seemed to accept me fully as her foster parent. In fact, she was all around more agreeable and placid than the owl had ever been. I worried that she would never become sufficiently aggressive—*mean*—to be able to survive in the wilds as an adult. Unfortunately, our agreeable relationship as well as the possibilities of her ever enjoying a free wild life were terminated by a severe storm, during which her shelter collapsed and injured her beyond recovery. I think I have never felt toward a bird the same sense of equality and understanding as I did with the eagle. I love birds as a species and have loved individuals dearly, but with the eagle I felt friendship; not slavish devotion, friendship.

While the eagle is almost certainly the largest wild bird I will ever be personally acquainted with, probably the smallest species I will ever know was a baby barn swallow. And if the eagle made me appreciate the lusty carnivorousness of the mightiest of the predators, nothing ever made me more fully recognize and appreciate the contribution birds make in our own fight against insects than attempting to find sufficient feed for that baby swallow. I say *our* because this event occurred after I had married and had children, and caring for the swallow ultimately required the efforts of the entire family and the contributions of understanding and helpful neighbors as well. The swallow was one of two nestlings delivered to our house by a small boy we knew. All matters being considered, I inclined to believe the barn swallows' mud nest had been unsettled by boy-play, but only if my own boyhood had been reproachless could I have summoned up a righteous indignation.

We agreed to do what we could with them and rigged up a box, with an imitation nest packed with cotton inside and heated by being placed on a heating pad. It certainly did not look in any way like a swallows' nest, but it did support their feeble bodies and prop up their heads. They had developed beyond that least appealing stage of nestling birds, in which they seem to be scarcely more than embryos or illustrations in a biology textbook. These were developing prickly little blue feathers, sticking from the skin as if they were designed to attach the skin to their nearly meatless frames. Their eyes were dark pools inside a membrane that had not yet separated to become pairs of eyelids. Their tiny beaks connected to saggy yellow flaps, like flaring but untidy riding breeches, where they joined the head. When they opened their beaks and these skin flaps stretched, the effect was almost as if the head had split into two hemispheres.

That open funnel into the maw of the tiny birds occupied our attentions during a good part of the daylight hours. At the approach of a hand the heads would open, extending and folding that yellow pouch which ended in a faintly pink void. Unfailingly, however, when we first touched them with whatever offering of food, the beak would snap closed. I hit on the expedient of using a toothpick, with a slightly split end to serve as a fork, for holding the food, reasoning that its form and feeling might suggest something of a parent bird's own beak. It seemed to work. Teased and caressed by the toothpick, the beak would remain open and the morsel would be zipped out of the forked end of the toothpick into the "mouth" and swallowed.

However, what to feed them quickly became a mounting problem, and an absorbing one it was. Again reason stepped in and suggested that since egg yolk was bound to consist almost exactly of what they were already formed from, if they could be persuaded to accept it, it should be healthful and nourishing. The idea was successful to a certain extent; each would take a few toothpick loads eagerly, but would continue to act hungry after they would no longer accept the yolk. Being swallows, we knew of course that their parents would have been supplying them with bugs, probably flying bugs exclusively.

Among the phrases which have probably never been spoken is, "You can never find a fly when you need one." We had a corral, horses, small livestock, but unaccountably we could not locate enough flies to satisfy those endlessly ravishing appetites. I bought fly swatters for all hands, plus spares for visitors, and we stalked the outdoors like War and Pestilence, hunting flies to feed the birds. We even smeared bacon drippings on our sidewalks and steps to draw flies. There were never enough. Sympathetic friends would drop by with small offerings in jars—pitifully small, if they only knew. Somehow we always ended up with enough flies to satisfy them and put them to sleep, but the hunts had to continue through their brief naps to rebuild a supply to feed them when they woke.

After caring for the babies for about a week, one of them died during the night for reasons we never knew. The other remained healthy, and its appetite seemed to expand to fill the void the death of its sibling had created. With four hunters in our family, plus outside volunteer help when it could be impressed, we were still barely able to supply the quantity of flies it required. No genius with statistics could appreciate as well as he who is catching them all by hand the quantity of flies four or five nestlings in the wild would require, nor the amount of insects a single swallow family removes from the environment in sustaining itself. When you are required to kill and capture part of that number of insects yourself, the statistics begin to mean something. A few swallows obliterate a whale of a lot of bugs.

As the wing feathers grew out in the little wings, the bird began to become more and more curious, insatiably curious about the world outside of its box. It would flutter to the edge and perch, for the most part reversing itself with care to cast its droppings outside of the nest. Shortly it began to fly. It was very cautious at first, and always tried to end its early attempts at flight back at the "nest." I rigged a coat hanger from a light fixture, adjusting it low above its box, and for a while it accepted this as a perch and point of vantage to rest and review the household.

Quickly it began to include the whole house in its flights, and particularly liked to perch and ride about on the rim of my glasses. It also began to enjoy family mealtimes, and, for all of its exclusive taste for flying insects at its own feedings, experimented happily with all kinds of human food. There was no keeping it off my plate, skittering about, sampling, testing, tasting. It discovered it had a particular taste for butter, peas, and fried chicken. As much as it seemed to enjoy fried chicken, this seemed to me to constitute a certain kind of outrage, tantamount to cannibalism, and I found myself getting almost gruff with him on the matter.

It was with some trepidation we allowed it its first experience outdoors. We were afraid that it might simply fly off, still somewhat inexperienced in flying, with no sense of self-protection, and no experience at all in what must be a most tricky business, catching flying insects on the wing. Our concern turned out to be wholly unnecessary. On its first venture outdoors, it would not fly at all. Finally I placed it on a clothesline; it made a beeline for my glasses and would budge no farther. On the second venture outdoors, it actually made a couple of short, tentative flights on its own, but invariably ended up retreating to the rim of my glasses. On the third day, I placed it on the clothesline, and, after a short period of reflection, suddenly it shot off across the yard, over the fence, and swooping low took off across the pasture. Finally it disappeared from sight, and after a few minutes, we were certain it had finally left us for good. But suddenly it appeared again, still flying low, boring toward me at full speed. He made a heavy landing on my glasses and hung there panting deeply, but obviously very pleased with himself.

On the next evening it immediately took off by itself, and when it returned after a few minutes, it was followed by a pair of adult barn swallows. When it swooped down to alight on my glasses, the adults circled close overhead, screeching most dolefully. "You fool! You fool!" they seemed to be saying. But he was no fool: he knew those birds were lying when they claimed to be swallows; genuine swallows, he knew for absolute certain, were much bigger, had red hair, and carried about with them a most convenient landing perch.

Naturally we wanted very much to establish a liaison between those adult swallows and our foundling, hoping they might at least pass on to him some helpful hints about catching bugs on the wing. Night after night, however, although it might fly a while with the wild swallows, always it came back to me, despite the wildest shrieks and pleadings of its natural kin. I was beginning to believe that ultimately it was going to be necessary, when the migration time arrived, to sit down with the bird over a set of maps and plot him his course to Guatemala. The wild vision even crossed my mind that he would be setting off with a minuscule map case and navigator's kit and an ever so small fly swatter.

We began making it spend its nights outside, and in the morning it would immediately fly into our house, cheeping most piteously. It could hardly contain itself until we had captured enough flies to feed it.

The final break was traumatic by necessity. We had been planning a week-long family trip, and clearly a full-grown swallow could not fit into our travel plans. And the response we got when we offered acquaintances a temporary job as a swallow sitter can best be imagined. It was well used to sleeping outside, however, and the season was advancing toward migration time, so reluctantly we departed, leaving the swallow outside to his own devices. When we returned, the bird was not to be seen.

The following spring an unusually fearless and inquisitive swallow showed up in our yard. It never landed on me, but flew around close by, and for a few days frequently perched on the clothesline when I was working in the yard. After a time I saw it no more, and assumed, whatever its origins, it had accepted its own categorical imperative and become one of the partners in a nesting enterprise. And very few humans, except me and my family, could truly appreciate the number of flying insects it would have to catch before it could again retire to winter quarters in the sunny South.

Of all the nestlings I have raised, the most jovial good fun was a young robin. We found him after a tumultuous thunder and wind storm, fluttering along a dirt road, apparently far from where he had been hatched. He was well feathered, but as yet completely flightless. He took to us immediately as foster parents and, after the swallow, had a refreshingly unfastidious diet. If we offered him egg yolk, he ate it; if we offered him bread soaked in milk, he ate it; if we offered him worms, he ate them. He found everything agreeable, and everything agreed with him. He liked everybody and everything, and his indomitable cheerfulness knew no limits. We named him "Percy."

After a week or so I took to feeding him worms exclusively, on the theory that, if he was going to make it on his own in the wilds, he should begin to think exclusively in terms of natural diet. A robin with a taste inclining toward bread and milk or hard-boiled egg yolks would find itself decidedly disadvantaged in its migrations. And happily worms are much easier to feed than flies, and it is infinitely easier to come by them in quantity. The tickle of a squirming worm against his chin would immediately trigger his mouth open, and with eager willingness, chomping, swallowing, and fluttering, the worm would quickly disappear. As large as worms are, it took a surprising quantity of them to appease his appetite. But then suddenly, after having ingested worms beyond count, he would close his beak tightly, blink his eyes a couple of times, utter a soft "cheep," and fall asleep.

As soon as Percy began to fly in short hops across the house, I took him outside to give him some lessons in self-feeding. The first time I sat him on the ground, he began to peck, and he had caught and eaten his own first bug before I ever saw what he was after. Thereafter we made available all opportunities for him to advance his cunning in finding his own food. Percy especially enjoyed gardening with me, and would watch with professional interest for the bugs I turned up or scared up. He would still huddle up and flutter his wings like a baby when I offered him a worm, long after he was perfectly capable of picking it up and feeding himself.

During this time we had a problem with our septic tank, and some of the happiest days Percy knew were spent dogging the plumber's footsteps as he dug out our pipes, feasting on the worms his shovel turned up. Our plumber was a gentle fellow, and fearing he would dump a shovelful of dirt on Percy's person, his caution probably added considerable expensive time to completing the job.

So permanently settled in our household was Percy that to resubmerge him in the life he was born to took a real effort. For the final weaning from us, we finally gave him over to the care of a sympathetic friend who lived some distance away, in hopes that he would gradually become separated from human companionship and begin to take up with birds. It seemed to work because he never demanded to come into the house of our friend and seemed to become less indiscriminately fearless of people. What happened to him ultimately we never knew, but when he took off he was able in all respects to feed and care for himself as well as if he had been reared by genuine birds.

If it ever happens, and it probably will, being the kind of person you are, that a bird foundling is dropped off on your doorstep, by all means make the effort to "bring it through." Keep it warm, *don't handle it* any more than is absolutely necessary, and try to approximate as closely as possible its natural diet. Don't be downcast if it doesn't survive; it certainly would not have survived without your efforts. You must be capable of looking the other way when an occasional bird dropping lands inconveniently. It won't last long, after all. Having brought up children is good training for raising baby birds. And when you finally succeed in finding the bird no longer needs you and can survive on its own, you can say good-by with a sense of pride and accomplishment. For the rest of your life you will have an appreciation of the part wild birds play in the natural scheme of things you could never have acquired otherwise.

GLOSSARY

Broiler factory: The system of raising fryers in many-tiered cages, by which means the largest number of chickens may be raised to eating size in the smallest amount of space.

Broiler mash: The commercially prepared feed given to chickens for the last four to six weeks before they are large enough to be eaten as fryers or broilers. The term is interchangeable with "finishing mash."

Brooder: An enclosure used for supplying warmth to baby poultry during their first weeks. A brooder may consist of a complete building, or a place within a building, or an enclosure in which warmth is supplied and a constant temperature maintained.

Glossary

Brooder house: The building in which baby poultry is kept during the first weeks when they must be supplied with heat. Sometimes the entire building is heated; sometimes the heat is supplied chiefly within a metal or improvised enclosure freestanding on the floor or hung umbrellalike from the ceiling.

Broody: The condition of a hen when she has "decided" she is in the mood to incubate a batch of eggs. A broody hen is characteristically crabby, and invariably clucks softly as she walks about.

Candling: The system of inspecting an egg by holding it against a strong light. The light will reveal, as shadows within the egg, if it is fertile, double yolked, or spoiled.

Capon: A male bird which has had its testes surgically removed at an early age. Capons grow larger and fatter than normal cockerels.

Caponize: The act of surgically removing the testes from a young male bird.

Clutch (of eggs): A nestful of eggs; a "setting" of eggs; a sufficient number of eggs for a hen to incubate.

Cockerel: A young male chick. Older birds are sometimes called "cockerels" by people who scruple at calling them "cocks."

Cracked corn: Corn which has been roughly milled, but not ground to a flourlike consistency. It is an old standby, general-purpose poultry feed.

Cull: A hen in a flock of layers that has been judged below standard in her egg production; also a bird rejected for any reason from a strain of breeders.

Culling: Removing from a flock unproductive or marginally productive laying hens; also removing from the flock birds judged to be inferior for any reason, particularly as breeding stock.

Debeaking: Removing the sharp pecking point of the beaks of chickens (or any other poultry) to discourage cannibalism.

Glossary

Droppings pit: A roosting arrangement for chickens that allows their feces to accumulate in an easily cleaned bin.

Dry-pick: To remove the feathers from a bird without first scalding in hot water.

Finishing mash: A feed given to developing chickens after they have outgrown their highest need for protein, to replace the more expensive starter mash.

Grit: Ground or crushed rock, usually granite, which is fed to poultry to provide the internal tools necessary for grinding and digesting their feed.

Hatchling: A bird, regardless of species, newly emerged from the egg.

Hover: A means of centralizing warmth for baby poultry; often an umbrellalike metal or improvised device hung or standing a few inches from the floor, with a heat source in its center. The word *brooder* is frequently used interchangeably with *hover*.

Incubate: To maintain eggs at a proper degree of warmth until a chick develops within and hatches.

Incubator: A mechanical device which artificially supplies warmth to eggs during the term necessary to turn fertile eggs into live baby birds.

Keelbone: The breastbone of a bird; in the front, it connects with the wishbone.

Keet: A baby Guinea.

Laying mash: A specially compounded commercial feed mixture fed to laying hens, alone, or in supplement with other feeds.

Lice: A parasitical insect which sometimes infests poultry.

Litter: Loose, absorbent material spread on the floors of brooder houses, chicken houses, etc. It may be straw, wood shavings, peat moss, ground corncobs, etc.—whatever is available and cheapest.

Glossary

Mites: An arachnidan (spiderlike) parasite which sometimes infests poultry.

Molt: To "shed" old feathers prior to growing a new coat, applied to birds of any species.

Pinfeathers: The emerging, undeveloped feathers of a young bird, or a bird just finishing its molt. When plucking a bird for the table, these are usually the most troublesome and difficult to remove.

Poult: The proper designation for a baby turkey.

Scald: To immerse a freshly killed bird quickly in hot water as an aid to loosening the feathers and make plucking easier.

Scratch (noun): A mixture of whole grains ground to various degrees of coarseness which is fed as a cheaper supplement to commercial starter and finishing mixtures. Sometimes scratch is fed exclusively through all stages of development, although when fed alone the rate of growth is sharply reduced.

Set (verb): As applied to poultry, to incubate a clutch of eggs, or to begin to incubate a clutch of eggs.

Setting (of eggs): The number of eggs one hen can incubate or set, or the number of eggs put into an artificial incubator at one time.

Squab: An immature pigeon at the age for eating; usually about a month old.

Starter mash: Commercially prepared high-protein feed which is fed to baby poultry during their first few weeks.

Unthrifty: A bird which is weak, sickly, or otherwise not worth the feed or attention necessary to keep or maintain.

Vent: The anus of a bird.

INDEX

Age, for butchering, 6
Agricultural colleges, poultry housing studies, 92
Antipecking medication, 86
Argentine pampas, 3
Artificial light
  for baby chick brooder, 14, 95
  for egg production, 39–40, 95–96
"As hatched" baby chicks, 21–22
Asia, 49
Augustus, 50
Aylesbury ducks, 70

Baby chicks
  adoption by broody hen, 57
  brooder (hover) for, 12, 13–14, 16
  certified disease free, 82, 83
  chick scratch for, 23–24
  cost of, 11–12, 22
  debeaking, 85
  disorders of, 84
  feeding equipment, 16, 17, 22–23
  feeding (for eggs), 37–39
  feeding (for meat), 22–25
  food consumption, 23
  housing, 11–13
  protein requirement, 22
  purchase sources, 21–22
  starter mash for, 22–23
  temperature for, 11, 15, 16
  watering equipment, 16–18, 82
  See also Day-old chicks
Bantams, 50, 51–54, 89, 103
  brooding, 53
  hatching and raising mallard ducks, 64
  hatching and raising turkeys, 74–75
  hovering other chicken breeds, 53, 54
  managing a brood, 57
Barley, as forage crop, 102
Barn swallow, caring for orphaned nestling, 118–22
Basement, raising chickens in, 10
Birds, caring for orphaned nestlings, 112–24
Blood clot in egg, 2
Boarders (nonlaying hens), 45
Breeding
  commercial standards, 61
  geese, 72
  program for, 60, 61
  selective, 60, 76
  turkeys, 74

Index

Breeds. *See* Chicken varieties
Broiler factory, 6, 19
Broiler mash, 24
Broiler ration, 24
Broilers, 10, 82
    feeding, 24
    housing, 92
Bronchitis, 83
Brooder (hover), 13, 16
    commercially made, 13–14, 15
    heat for, 14
    homemade, 14
Brooders (brooding hens), 53
    coop for, 55, 56
    eggs for, 55
Broodiness, 54
Buff Orpingtons, 35, 59
Butchering, 26–27, 30–31
    age for, 6
    commercial shop for, 26
    ducks, 68
    fryers, 6, 26, 106
    old-fashioned way, 26–27
    scientific way, 27
    turkeys, 75, 76

Calcium requirement, 39
Candling eggs, 2
Cannibalism, 75, 84–86
Caponizing, 25–26
    kits for, 25
Capons, 25, 26
Carving, 30
Casting, 117
Catalytic heaters, 15
Cats, 10, 30, 89, 103, 107
Cattle, 33
Certified disease-free chicks, 82, 83
Chicken houses, 12, 13, 91–103
    cleaning, 82, 93
    droppings pit in, 82, 96–97
    electricity for, 95–96
    feeder attached to, 93
    feeding arrangements, 97–98
    heating, 13–16, 95, 96
    nests, 93, 97, 98, 99
    open shelter, 93
    prefabricated, 92–93
    sun-porch addition, 99, 102
    waterer attached to, 93
    water for, 16–18, 95
    See also Housing
Chicken keeping, 63
    economics of, 107–8, 109
    for eggs, 2–5, 33–47
    local ordinances against, 10–11
    for meat, 2, 5–6, 9–31
    number in flock. *See* Flock size
    reasons for, 2, 6
    space for, 10–20
Chicken manure, 108–9
Chickens
    ancestry of, 49
    controls for predators, 86–90
    superior flavor of homegrown, 106
    See also Chicken varieties
Chicken varieties (breeds), 6–7, 58, 59
    Buff Orpingtons, 35, 59
    crossbreeds, 21
    dual purpose (combined egg and meat production), 21, 35, 59
    for eggs, 21, 35–36
    exotic, 6–7, 21
    game, 50, 51
    Hubbards, 21, 35, 59, 106
    indigenous to region, 59–60
    for meat, 21, 35
    New Hampshires, 59
    ordering fertile eggs for new strains, 61
    Plymouth Rocks, 21, 35, 59
    Rhode Island Reds, 21, 35, 51, 53, 59
    White Leghorns, 21, 35–36, 40, 42, 45, 97, 103
    White Rocks, 21, 35, 59
Chicken wire
    for fencing, 102
    for housing, 93
Chick scratch, 23–24
China, 70
Cleanliness, 12, 15
    disease preventive, 82
    droppings pit, 82, 96–97

Index

factor in chicken house selection, 93–94
Clucking, 54
Cockerels, cocks, 25, 36
  caponizing, 25–26
  sex designated in day-old chicks, 21, 36
  See also Roosters
Cockfighting, 50
Compost, 108
Costs
  baby chicks, 11–12, 22
  brooders, 14
  chicken house, 94, 95
  cutting, 5
  feed, 22, 23, 38
  guineas, 79
  pullets, 36–37
County extension agent, questions for
  about breeds that are successful in area, 21, 59
  about cannibalism, 85
  about caponizing, 25
  about chicken houses, 92
  about forage crops, 102
  about rat control, 87
  about vaccination, 83
Coyote control, 89
Cracked corn, 24, 38
Culling, 45
Culls, 45, 47

Dampness, protection from
  baby chicks, 16, 18, 82
  ducklings, 67, 73
  goslings, 73
Darwin, Charles, 3
Darwin's rhea, 3
Day-old chicks, 11–12
  as hatched (hatchery run), 21–22, 36
  sex-designated, 21, 36
  See also Baby chicks
Deacon's nose, 29
Debeaking, 85
Disease control, 11, 81–83
  certified disease-free chicks for, 82, 83
Dogs, 10, 89, 103
Drake, 65, 71
Dressing, 6, 28–29, 30–31
  ducks, 69
  New York dress, 69
Droppings pit, 82, 96–97
Ducks and ducklings, 63–71
  Aylesbury, 70
  butchering, 68
  dressing, 69
  eggs, 69, 70–71
  feeding, 66, 68
  hanging, 69
  Khaki Campbell, 70–71
  mallard, 63–64
    tame, 70
  meat producers, 70
  monogamy of, 71
  Muscovy, 70
  number in flock, 71
  Pekin, 64, 70
  plucking, 68–69
  roasting size, 67
  Rouen, 70
  White Pekin, 70

Easter eggs, 69
Egg production, 33–47
  artificial light for, 39–40, 95–96
  bantams, 52
  combined with meat production, 21, 35
  ducks, 70–71
  establishing a laying flock, 21, 36–37
  feeding for, 37–39, 40–41, 42
  housing for laying hens, 92, 96
  laying-condition indicators, 45–47
  laying mash, 37, 39
  nonlaying hens, 45
  pullets, 21, 36–37, 43
  White Leghorns for, 21, 35–36, 53
Eggs
  blood clot in, 2
  candling, 2
  dark- and white-shelled, 35
  double yolk, 43

Index

duck, 69–70, 71
farm fresh, 3–5
fertile, 2, 7
flavor, 2, 4–5
freezing, 43–44, 70
fresh, 2–3
gathering, 3, 7, 41, 107
guinea, 79
hung up in hen, 86
incubating, 61
ordering for new strains, 61
pullets', 43
for setting, 55
storing, 43, 44
surplus, 105, 107
in water glass, 44
Eggshells, 39
dark and white, 35
Egg white, judging freshness by, 4
Egg yolk
double, 43
judging freshness by, 4
Electricity, for chicken house, 15, 95–96
Embden geese, 72, 73

Fantail pigeons, 77
Feather removal. *See* Plucking
Feed, feeding
baby chicks, 22–25, 37–39
broiler mash, 24
broiler ration, 24
chick scratch, 23–24
cost, 22, 23
ducks, 66, 68
for egg production, 3–4, 37–39, 40–41, 42
finishing mash, 24
finishing ration, 38
forage crops, 102
foraging, 39, 64, 67, 68, 71, 73, 76, 102
fryers, 24
garden- and kitchen-scrap supplement, 41, 42
geese, 71, 73
grain, 38

growing mash, 38, 68, 75
growing ration, 38
for meat production, 5–6, 22–25
nutritional-deficiency-disorder prevention, 82
roasters, 24–25
starter mash, 22–23, 38
starter pellets, 68
turkeys, 75
Feeding equipment, feeders, 97–98
attached to open shelter, 93
baby chicks, 16, 17, 22–23, 25
built for specific need, 25
fly-up, 97–98, 100
gravity-fed, 97
homemade, 17
metal, 16
overhead (suspended), 40, 97
wooden, 16, 25
Fencing, outdoor runs, 102–3
Fertility, 2, 7
Fertilizers
chicken manure, 108–9
3-8-7 commercial mixture, 108
Field owl, caring for orphaned nestling, 113–16, 117
Finishing mash, 24
Finishing ration, 38
Flavor
eggs, 2, 4–5
meat, 5–6, 24–25, 106
Flock size
baby chicks purchased, 22–23
ducks, 71
egg production, 40, 41, 107
factor/function determinant, 39
family size, 41, 105–6
fryers, 11–12, 20, 106
Fly-up feeders, 97–98, 100
Forage crops, 102
Foraging
chickens, 39
ducks, 64, 67, 68
geese, 71, 73
outdoor runs for, 102
small-grain planting for, 102
turkeys, 76

Index

Fowl pox, 83
Freezing
  eggs, 43–44, 70
  fryers, 30
  roasters, 30
Fresh and farm-fresh eggs, 3–4
  flavor, 4–5
  white indicator, 4
  yolk indicator, 4
Fryers, 6, 10, 19, 20, 39
  annual yield in pounds, 106
  breeds for, 21
  butchering, 6, 26, 106
  cost compared with retail prices, 106
  dressed weight, 20, 106
  feeding, 24
  food consumption per pound of meat produced, 24
  freezing, 30
  housing, 92
  number in flock, 11–12, 20, 106
  selling surplus, 105

Galen, 7
Gall bladder removal, 29
Game chickens, 50, 51
Game cock, 50
Garage, raising chickens in, 12, 13, 20, 92
Garden scraps, as food supplement, 41, 75
Geese and goslings, 12, 30, 54, 71–73
  breeding and mating, 72
  Embden, 72, 73
  feeding, 71, 73
  foraging, 71, 73
  plucking, 68
  Toulouse, 72–73
  weeding vegetable crops by, 71–72
  White Chinese, 73
Giant Homer pigeons, 77
Giblets, 29
  freezing, 30
Gizzard, 29
Golden eagle, caring for orphaned nestling, 116–18

Goslings. *See* Geese and goslings
Grain, 38
Grit, 23, 24, 39, 75
  combination hopper for crushed oystershell and, 38
Ground owl, 114
Growing mash, 38
  for ducks, 68
  for turkeys, 75
Growing ration, 38
Guineas, 54, 77–79, 89, 103

Hanging (ducks), 69
Hatchery run chicks, 36
Hatchlings. *See* Baby chicks; Day-old chicks
Hawk control, 88–89
Hawthorne, Nathaniel
  The House of Seven Gables by, 60
Heart, 29
Heat, heating, 13–16, 95, 96
  brooder, 13–14, 15
  catalytic, 15
  electricity, 15, 95, 96
  infrared lamp, 12, 13
  kerosene, 15
Hens, 7
  adoption of baby chicks by, 57
  broody, 53, 54–56
  clucking, 54
  contented, 58
  culls, 45
  egg hung up in, 86
  establishing laying flock, 36–37
  laying, 20, 37, 42–43, 45, 92, 96
  laying-condition indicators, 45–47
  laying-performance recording, 60
  molting, 44–45
  nonlaying (boarders), 45
  record keeping for breeding program, 60
  replacement flocks, 44
Hogs, 33–34
Hopper, combination grit and oystershell, 38
Horses, 33
Hot-wax plucking, 68–69

Index

House of Seven Gables, The
   (Hawthorne), 60
Housing, 11–13, 91–103
  baby-chick brooder, 12, 13–14, 16
  basement, 10
  broody hens, 55, 56
  enclosed porch, 10
  garage, 12, 13, 20, 92
  laying hens, 92, 96
  open shelter, 93
  outside runs, 102–3
  pigeons, 76–77
  prefabricated, 92–93
  range shelter, 19
  turkeys, 76
  See also Chicken houses
Hover. *See* Brooder
Hubbards, 21, 35, 59, 106

Immersion heater, for waterer, 99
Incubation, 61, 62
Incubators, 61–62
  electric, 66
  table-top, 61
Infrared lamp, 12, 13
Internal parasites, 2, 82

Keets, 79
Kerosene heaters, 15
Khaki Campbell ducks, 70–71

Lard, storing eggs in, 44
Laryngotracheitis, 83
Lawn fertilizer, 108
Laying-condition indicators, 45–47
Laying flock, 36–37
Laying mash, 37, 39
Lice, control of, 83–84
Light bulbs, 14
Limestone, crushed, 39
Litter, 15, 67, 82
Liver, 29

Magpies, 89
Mallard ducks, 63–64
  tame, 70
Mangle (food supplement), 42

Mark Antony, 50
Mating
  geese, 72
  turkeys, 74
Meat production, 2, 5–6, 9–31
  breeds for, 21, 35
  combined with egg production, 35
  ducks, 70
  feeding for, 5–6, 22–25
  squabs, 77
  turkeys, 75–76
  See also Butchering; Fryers;
    Roasters
Middle Ages, pigeon breeding in, 77
Millet, 38
Mites, control of, 83–84
Molting, 44–45
Monogamy
  ducks, 71
  geese, 72
Muscovy ducks, 70

National Bureau of Standards, 90
Nesting, 52–53
Nestlings, caring for wild orphans,
  112–24
Nests, 97, 98, 99
  attached to open shelter, 93
  trap, 60
Newcastle disease, 83
New Hampshires, 59
New York dress (ducks), 69
Nitrogen, 108
Nouns of assemblage, 30

Oats, as forage crop, 102
Outdoor run, 99, 102–3
  fencing for, 102–3
  forage crops for, 102
Overhead (suspended) feeder, 40, 97
Owls
  field, 113–16, 117
  ground, 114
Oystershell, crushed, 39
  combination hopper for grits and,
    38

Index

Paralysis, 84
Peas, as forage crop, 102
Pekin ducks, 64, 70
Phosphorus, 108
Pigeons, 76–77
  Fantail, 77
  Giant Homer, 77
  Pouter, 77
  rock, 77
  Roller, 77
  Tumbler, 77
  White King, 77
Plucking (picking), 27–28
  ducks, 68–69
  melted-wax method, 68–69
  scalding method, 27–28
  turkeys, 76
Plutarch, 50
Plymouth Rocks, 21, 35, 59
Porch, raising chickens on, 10
Potash, 108
Poults, 74
Pouter pigeons, 77
Predator control, 86–90
Protein
  baby chick requirement, 22
  in starter mash, 75
  turkey requirement, 75
Pullets
  baby-chick sex designation for, 21, 36
  cost, 36–37
  establishing a laying flock with, 36–37
  first eggs laid by, 43
  ready-to-lay, 37
  started, 36–37
  vaccination, 83
"Pynchon Chickens," 60

Rabbits, 12, 30
Range feeding, 39. *See also* Foraging
Range shelter, 19
Rat control, 86–87, 103
Record keeping
  breeding program, 60
  hens' laying performance, 60

Respiratory ailments, 83
Rheas
  Darwin's, 3
  greater, 3
Rhode Island Reds, 21, 35, 51, 53, 59
Roasters, 7–8, 20, 82
  breeds for, 21
  butchering weight, 24–25
  capons, 26
  dressed weight, 106
  ducks, 67
  feeding, 24
  freezing, 30
  per-bird space requirement, 19
  selling surplus, 105
Robin, caring for orphaned nestling, 122–24
Rock pigeons, 77
Roosters, 7–8
  fighting instincts, 50
  hawk warnings by, 88
  See also Cockerels
Roosts, 93, 96
Rouen ducks, 70
Rye, as forage crop, 102

Scalding, for plucking, 27–28, 68
Setting a hen, 54–56
Sex-designated baby chicks, 21, 36
Skunk control, 87–88, 103
Space, 10–20
  per-bird requirement, 19
Squabs, 77
Starter mash, 22–23, 38
  for turkeys, 75
Starter pellets, for ducks, 68
Sunflowers, 42
Storing eggs, 43
  warm-lard method, 44
  water-glass method, 44
Swans, 54

Table scraps, as food supplement, 41, 65, 66, 75
Tame mallard ducks, 70
Taste, superiority of homegrown chickens and eggs, 5, 106

Index

Temperature, 11
  for baby chicks, 11, 15, 16
  brooder, 13–14, 15
  for ducklings, 67
Terms of address, 30
Tiring, 117
Toulouse geese, 72–73
Trap nests, 60
Traps, for poultry predators, 86, 87
Tumbler pigeons, 77
Turkeys, 73–76, 89
  butchering, 75, 76
  commercially bred, 74
  stampeding by, 76
Twain, Mark, 26, 51

United States Department of Agriculture
  caponizing bulletin, 25
  chicken-nutrition bulletins, 22

Vaccination, 82, 83
Varieties
  chicken. *See* Chicken varieties
  ducks, 64, 70–71
  geese, 72–73
  pigeons, 77
  turkeys, 76
Vegetation, as food supplement, 41–42, 64, 75
  See also Foraging
Vermin control, 11, 83–84

Water, watering, 82
  baby chicks, 16–18
  in chicken house, 95, 99
  ducks, 66, 68
  geese, 72
  See also Dampness, protection from
Waterer
  attached to open shelter, 93
  cleaning, 82
  gravity fed, 18, 98
  immersion heater for, 99
  portable, 99
Water-glass egg storage, 44
Weasel control, 87, 103
Wheat
  as feed, 38
  as forage crop, 102
White Chinese geese, 73
White King pigeons, 77
White Leghorns, 21, 35–36, 45, 97, 103
  disposition, 36
  egg-production capacity, 40
  fencing for, 36, 103
  food requirement, 42
White Pekin ducks, 70
White Rocks, 21, 35, 59
Wind duck, 69
Worm problems, 83

300